D0356228

Dr. Eleanor's
BOOK OF
Common Spiders

WITHDRAWN FROM COLLECTION
VPL

DR. ELEANOR'S

BOOK OF

Common Spiders

※

CHRISTOPHER M. BUDDLE

and ELEANOR SPICER RICE

THE UNIVERSITY OF CHICAGO PRESS
Chicago and London

The University of Chicago Press, Chicago 60637
The University of Chicago Press, Ltd., London
© 2018 by The University of Chicago
All rights reserved. No part of this book may be used or repro-
duced in any manner whatsoever without written permission, ex-
cept in the case of brief quotations in critical articles and reviews.
For more information, contact the University of Chicago Press,
1427 E. 60th St., Chicago, IL 60637.
Published 2018
Printed in Canada

27 26 25 24 23 22 21 20 19 18   1 2 3 4 5

ISBN-13: 978-0-226-33225-3 (paper)
ISBN-13: 978-0-226-33239-0 (e-book)
DOI: 10.7208/chicago/9780226332390.001.0001

LCCN: 2017026799

♾ This paper meets the requirements of ANSI/NISO
Z39.48–1992 (Permanence of Paper).

CONTENTS

INTRODUCTION: WHAT IS A SPIDER?

When we think about earth's animals, most of us think of creatures like dogs or elephants or birds. However, the majority of earth's animals don't have big, floppy ears or feathery tails. Most animals on earth have hard shells, called exoskeletons. Instead of paws, they have many legs extending outward from their bodies, swinging from joints. Instead of having one heart, an appendix on one side, and a bile duct on the other, their bodies and organs are symmetrical, so if we cut one right down the middle, the left half will perfectly mirror the right half. Most animals on earth, including spiders, are arthropods.

Other arthropods include crabs and lobsters, millipedes and centipedes, grasshoppers, butterflies, and beetles. Although it might be obvious that some arthropods are not very closely related (lobsters and butterflies, for example), many might look alike to us. Sometimes people think spiders are "creepy." They keep their distance, and lump spiders together with insects that they find similarly shiver-inducing. But spiders are not insects. They are part of a completely different group called arachnids (scientific class Arachnida). After a closer look, the differences between spiders and insects are clear.

Insects have three body segments: a head with a brain and two antennae, a thorax with six legs and wings, and an abdomen that holds most of their organs. But spiders, like all other arachnids, have only two body segments, and they never have wings.

2 body parts + 8 legs – wings = spider

The head and thorax of an arachnid is combined into a "cephalothorax," from which eight legs extend. An arachnid's abdomen is similar in form and function to an insect's abdomen. Many life forms fit the arachnid mold, including scorpions, mites, daddy long legs, and ticks.

Within the class Arachnida, spiders are in the scientific order Araneae. All members of the Araneae order have four pairs of legs, silk-spinning organs called spinnerets on their abdomens, and big, chunky "jaws" called chelicerae, tipped with venom-injecting fangs. They also have two short, leg-like extensions on the front of their heads called pedipalps ("palps" for short), which they use for any-thing from moving objects around to mating, to tasting and smell-ing, to "hearing" vibrations. A skinny waist, called a pedicel, con-nects the abdomen to the cephalothorax. Consider the blob-like tick or the chunky, long scorpion for comparison. Because they lack the Araneae pedicel, their waists can be much harder to find!

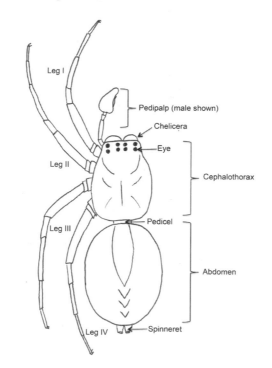

The anatomy of a spider.

Leg I

Pedipalp (male shown)

Chelicera

Eye

Leg II

Cephalothorax

Leg III

Pedicel

Abdomen

Leg IV

Spinneret

Growing Up, Spider-Style

Like many other baby arthropods, baby spiders, called spiderlings, emerge from eggs. When a spiderling first makes an appearance, it looks like a miniature version of its parents. To grow, it must shed its tough, tight exoskeleton (a process called molting), revealing a larger spider underneath. Spiders continue to molt until they reach adulthood. Their final molt signifies maturity, and after that molt they can mate and produce spiderlings of their own. Spider lifespans range from a single season to more than a decade. And depending on the species, it may take a long time to reach adulthood.

Male and female spiders often look too similar to tell apart until that last molt. However, once they shed their last childlike skins, a little training can help us tell the two sexes apart. A male spider's pedipalps have enlarged ends that, at first glance, resemble boxing gloves. Instead of buffering punches, these gloves store the sperm a male spider produces in his abdomen. A female spider's pedipalps remain slim and trim. On her abdominal underside, she has a sperm receptacle called an epigynum

TOP Just emerged from their egg sac, these common house spiderlings hang out in the web for a while. *Photo courtesy of Matt Bertone.*

BOTTOM In order to grow, spiders shed their tight outer skins, a process called molting. *Photo courtesy of Sean McCann.*

that fits like a perfect puzzle piece with her lover's puffy palps. Male and female spiders can differ in size, shape, and color. In many species, male and female spiders hardly resemble each other and can seem like different species altogether. For several species, we rarely see males at all.

As is the case for these western black widows, female spiders can look quite different from their male counterparts. *Photo courtesy of Sean McCann.*

Masterful Weavers

How does an orb-web spider build its perfect, spiral web between two distant objects? Orb-web construction is one of nature's marvels. First, the spider climbs a blade of grass, tree branch, porch railing, or some other elevated structure and raises its abdomen to release a strand of silk into the breeze. The spider then releases strand after strand until one snags on an object some distance away. Once the spider makes contact across the void, she pulls her silk tight to establish a bridge. Then, she walks along that bridge, placing reinforcements along the first strand of silk as she goes.

Near the center of the bridge, she drops down, releasing silk and pulling down strands from the main bridge. When gravity has done its part, and a vertical triangular shape has formed, the spider releases a vertical strand from the bottom point of the triangle and fastens it to an anchor point below.

With this Y framework in place, our crafty constructor travels back up to the center and along a short side of the triangle, then drops down to another

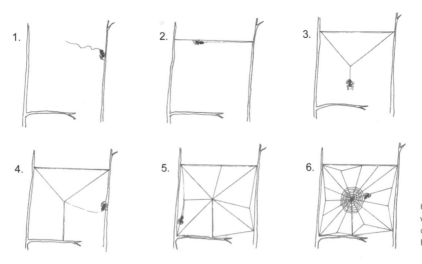

How an orb-web spider constructs her web.

anchor point. She repeats this process until all the radius threads are in place, making a scaffolding for her spiral.

Next, she travels to her web's center and lays down a temporary, outwardly expanding spiral, measuring the distance with her legs. Once done, she reverses direction, pulling up the temporary spiral and this time laying down a final, more densely packed sticky spiral, all the way back to the central hub. With a ready web, she waits for meals and lovers to stop by.

Many animals, from crickets to ants to caterpillars, beetles, and fleas, produce silk, but spiders are the true masters of silk production and use. Arachnologists use spider webs to help place spiders into groups of relatives. For example, orb-weaving spiders build vertical, netlike spiral webs to snare flying insects. A closely related group, the long-jawed orb weavers (page 19), build horizontal webs with a central opening to capture insects like mayflies as they emerge from the water. Cobweb spiders build chaotic, almost haphazard cobwebs in basement corners or tree cavities, tangling unsuspecting prey. And in the early morning we can see silver dew

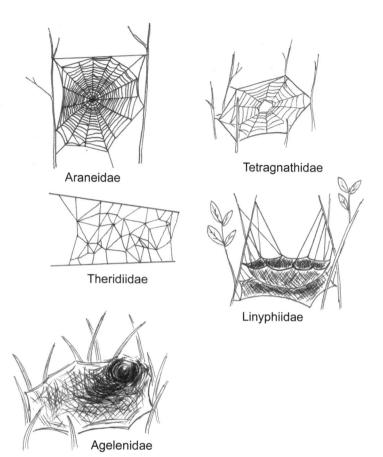

Araneidae

Tetragnathidae

Theridiidae

Linyphiidae

Agelenidae

Different spider families construct different types of webs, and some spider families do not construct webs at all.

collected on the flat webs of grass spiders (page 26), with their silken spider retreats disappearing into our lawns. It's easy to identify spiders based on their web types.

Some spider species don't build silken webs to capture prey, but they still spin silk for other purposes. For example, some species use "dragline" silk as safety lines, tethering themselves to tall objects and catching themselves if they fall. Many species wrap their prey in silk, immobilizing struggling animals to make for easier meals.

Spiderlings cast silken strands in the wind and, in a process called ballooning, use these lines to sail through the air and disperse to new locations. Spiders also use silk to construct protective cases for their eggs, called egg sacs.

But spider silk isn't just for spiders: humans harness spider silk technology for engineering and medical uses. Because spider silk is both stretchy and strong, engineers attempt to mimic its properties when synthesizing new materials for building and clothing. Some new kinds of body armor also use spider-silk technology— its strength protects, and its flexibility allows the wearer to run and jump freely. The medical industry is exploring spider silk's possible antiseptic properties as well as its potential for wound care.

Eyes on the Prize

With the exception of a few blind species, most North American spiders have eight eyes. You can tell a lot about a spider by looking it in the eye (or eyes). For example, big-eyed spiders, like jumping spiders, usually live their lives in the daytime and out of webs. They need those enlarged peepers to hunt prey. Spiders who while away most of their hours in webs, like writing spiders, tend to use their sense of touch more than their sight to mate or capture their prey, feeling the vibrations of struggling organisms caught in the web or the tentative strums of a potential lover waiting webside. These spiders have much smaller eyes. Other spiders, like cave spiders, live in total darkness and do not need to see anything at all. Their eyes are very small or even absent.

Arachnologists use spider eye size and arrangement to help assemble spiders into related groups. For example, wolf spiders have two big eyes in the middle of their heads, with a row of four smaller eyes beneath them. Farther back on their heads rest their remaining two eyes. Jumping spiders, on the other hand, have two huge, cartoon-like eyes right in the front of their faces, which they use to

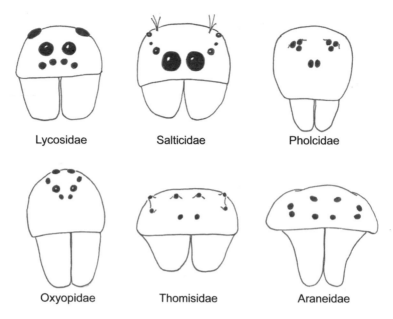

Lycosidae   Salticidae   Pholcidae

Oxyopidae   Thomisidae   Araneidae

Eye arrangements for some of the spider species featured in this book. Use this as a quick guide to help figure out what species you might have in your basement, garden, or local forest.

spot their meals. The rest of their eyes ring their heads like a garland, instead of being stacked like wolf spider eyes.

What Is Common?

We titled this book *Dr. Eleanor's Book of Common Spiders*, but how did we define what is common? Some species might seem rare only because they are hard to find, such as the purse-web spider (page 71). And some species might seem common just because they show up in the news a lot or because people talk about them over the dinner table or in schoolyards, such as widow spiders (page 29). We certainly don't know enough about spiders to draw good range maps, and there are no online resources to track species distribution. The bottom line is that even arachnologists don't know how to properly determine which spiders are truly common.

Despite this, we have opted to tell stories about the spiders that are common enough to be stumbled upon with some regularity, either in a local forest, behind your toilet, building a web across your front door, or scurrying along a brick wall near the bus stop. Some of the species we discuss are common only in some parts of North America, but even the East Coasters will have relatives with similar biology living out in California. And although we wish to tell the story of ALL the thousands of spider species in North America, we had to pick and choose. In addition to the more detailed chapters on thirteen common (plus one not-so-common) species, we have also included sidebars about twelve other species you might encounter in North America. We hope all these stories will inspire you to look at spiders a little differently and appreciate them for their marvelous life history.

01 WRITING SPIDER

SPECIES NAME: *Argiope aurantia*

AKA: golden writing spider, golden orb weaver, black and yellow Argiope, golden garden spider

BODY SIZE: females: 0.7–1.1 inches; males 0.2–0.3 inches

WHERE THEY LIVE: Writing spiders live in temperate areas of North America and Central America, from southern Canada to the United States and south from Mexico to Costa Rica. They can be found but are not common in the Rocky Mountains and the Great Basin.

Some say the garden spider's stabilimentum resembles handwriting. *Photo courtesy of Joe Lapp.*

In the Southern United States, legend has it that if you see your name written in a writing spider's web, you'll be the next to die. Every summer and fall, when I was a child, I would lean over the side of our porch, between the boxwood and the water meter, where our writing spiders set up shop. I would peer into the zigzag zippering the middle of these webs, terrified I'd see my name or, worse, the name of one of my parents or my brother. Each year, I would feel very sorry for Mr. ZZZZZZ or Mrs. NNNNNN, depending on which way I turned my head.

But the writing spider has more interesting talents than fortune telling. Its web zipper has a real name, stabilimentum, and perhaps works better for warning birds not to fly into the writing spider's expansive, sticky web or for luring prey than it does for predicting which funeral we'll be attending next. By the time you and I see the big writing spiders gently rocking on their webs in our gardens, they have already survived months, growing up slowly.

If you want to get up close and personal with a spider, writing spiders are among the best candidates for the job. For one, they're easy to see because they're huge and brightly colored. One of the largest of the North American orb weavers, the female writing spider's body can spread to more than one inch long. Also, they're beautiful. While we don't often see the smaller males, females have silvery hairs covering their backs, and their heads shine like dew in the morning sun. Their big, black, eye-shaped abdomens are blotted with yellow or pale spots and globs, making them more like living paintings than garden predators.

Unlike jumping spiders or other webless wanderers, golden writing spiders build giant, vertical, dreamcatcher-like webs, stretching their silk between tall grass and shrubs, in gardens or backyards. These webs help make writing spiders easy to watch. Writing spider females have three claws per foot. They use that third claw for handling their silk while spinning their webs. Down the center of these nets zigs and zags the stabilimentum. Females usually perch in the

This Argiope species' zigzag stabilimentum is woven in a cross shape. *Photo courtesy of Dinesh Rao.*

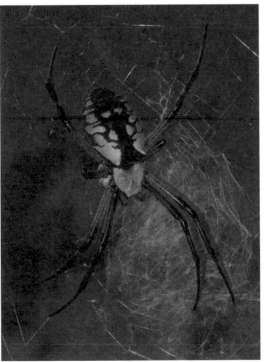

Some of North America's largest common spiders, garden spiders are also among the most harmless. *Photo courtesy of Maxim Larrivee.*

Garden spiders string their webs between our garden plants. *Photo courtesy of Sean McCann.*

center of their webs, halfway down the stabilimentum. They hold their long legs in pairs to form an X, and rest head-down in their webs as they gently rock back and forth in the warm, late summer breeze.

At the end of their adult lives, in the autumn months, writing spiders lay their egg cases, tucked into a safe area of the garden near their webs. Then they die, leaving those cases to overwinter with no protection. By spring, tiny spiderlings emerge and strike out on their own. These small spiderlings often escape our notice as they move about our lawns, capturing what tiny prey they can, such as gnats and flies. As they grow, so do their webs, and by the time we start paying attention to them, they're capturing much larger prey like moths, butterflies, beetles, and flies. These insect buffets are noticed by other critters, too. Some opportunistic hunting wasps, for example, see the struggling meals in writing spider webs and deftly snatch these snacks for their own.

While adult female writing spiders pass their days in their webs, capturing and devouring prey, males wander around, hoping to find a mate. Because females perceive anything tangled in their webs as a delicious meal, males must approach their prospective paramours with caution. Like a hopeless romantic, a male will tap out a little love song on the female's web. If the female likes what she's feeling, she'll let her new boyfriend approach her.

In the animal world, biologists often measure an individual's success by its ability to reproduce. Finding a mate is one step toward reproduction, but the act of mating doesn't by itself always guarantee an individual will produce offspring. For example, say a male spider successfully mates with his spider girlfriend and then wanders on off into the world, leaving her to hang out in her web alone. She may well accept another male and mate with him. His sperm would mix with the first male's, reducing the number of eggs that the first spider male's sperm will fertilize. The second spider might even scoop out the first spider's sperm, eliminating the first spider's chance at reproductive success.

Yes, it's a tricky world out there for spider lovers of all kinds. But for male writing spiders, after doing all that hokey pokey to avoid getting eaten by potential girlfriends and whatnot, it's especially important to do all they can to increase the chances that they will father offspring of their own. Writing spider males literally give it all they have: after mating, they die.

Here's what happens. First the male taps out his love song and gets accepted as a mate. Then, he inserts his pedipalps into the female and deposits his sperm. As soon as all his sperm gets into the female's body, the male dies, and his body hangs there from the poor female until it is eaten or falls off. His dead body acts as a "mating plug," blocking the female from mating with other males until she lays the eggs he fertilized.

It's hard to imagine the writing spider's weeks of wandering as a tiny creature, its beef with wasps, or its tragic sex life just by glancing at a female's tranquil sway in the afternoon sun. But maybe, if you have the time, you can do more than glance. Take a minute to watch her pick across the web with those silk-handling toes. Look for an egg case tucked beneath some leaves, or maybe inspect her diet by surveying who's dangling in her web. If you're lucky, you might even spot a successful lover, hanging blissfully dead after a job well-done.

Bolus spiders skip the traditional web net and instead have a cowboy-like way of fishing for prey. They spin a lump, or bolus, of silk with a gob of sticky wax on one end and a rope-like tether on the other. All while mimicking the smell

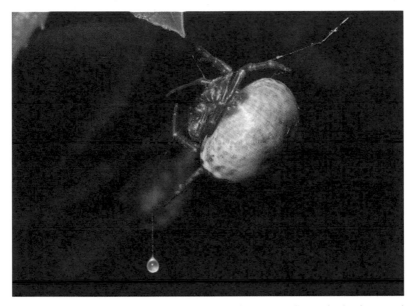

The bolus spider swings its sticky silk ball to capture prey. *Photo courtesy of Matt Coors.*

of female moth pheromones—those odors that put male moths in the mood for love—they swing their bolus around like a tether ball of doom. As it spins, love-hopeful moths fly in to check out the female moth scent. These lured males get whapped by the sticky bolus and reeled in to become bolus spider supper.

02 BARN SPIDER

SPECIES NAME: *Neoscona crucifera*

COMMON NAME: barn spider

BODY SIZE: 0.35–0.75 inches

WHERE THEY LIVE: Barn spiders live along the East Coast, spreading south and west to Arizona and Southern California. They build their webs in natural areas like forests as well as in urban and suburban areas, stretching their silk between human-built structures like fence posts, streetlamps, and back porches.

The Native American Ojibwa people (called Chippewa by others) have a story that a spider woman, Asibikaashi, protected them. As the Ojibwa nation spread across North America, Asibikaashi had a hard time trekking around that great distance to take care of everybody. To help Asibikaashi protect the children, Ojibwa mothers and grandmothers wove webs from plant fibers or yarn in willow hoops and placed them near their children's beds. Each night, dreamcatchers trapped harmful and bad dreams but allowed the good dreams to pass through to the children. In the morning, everything snagged in the dreamcatcher webs disappeared. Barn spiders are nature's Ojibwa dreamcatchers.

Barn spiders tiptoe along webs stretched from Ontario to Florida, and from the eastern United States to California. With adult females just under an inch long, they look like large, colorful acorns crouched in their webs' hubs. No two barn spiders look alike. They come in every shade of orange and brown. Some have beautiful patterns extending from a light mark down the center of their abdomens, while others have no pattern at all.

Like the dreamcatchers, our barn spiders work best at night. They

A barn spider waits for its prey. *Photo courtesy of Matt Bertone.*

stretch their vertical, netlike webs between plants, house eaves, or even in the crook of a "no parking" sign. Their impressive webs can span two feet or more. Barn spiders use two types of silk, produced by two different silk glands, to build their webs.

The first type of silk, which forms the support frame of the web, is called major ampullate silk. Major ampullate silk is built for strength. Five times stronger than steel when it's humid out, it keeps the web spread out and proud. The other type of silk, called viscid silk, forms the intricate, dreamcatcher-like net in the web's center. This silk, though not as strong as the web's frame, is highly elastic. That way, when a determined moth comes barreling toward the web at full speed, it doesn't break through. Instead it just boings into the web like a trampoline. A very sticky trampoline. When barn spiders lay down their stretchy viscid silk lines, they squeeze little droplets of glue in the spots where they want to net their fast food.

And boy, can they grab some grub. While dreamcatchers catch

bad dreams each night, barn spiders go more for bugs. Barn spiders build their nests near lights or the moon-exposed areas of forests so they can hook all the hopeful flying insects like beetles, moths, and flies that dart around bright lights. With their champion meshwork, these spiders often capture so many insects that they are considered "wasteful feeders," catching far more prey than they can actually eat.

Our orb-weaving friends like to stay close to us, and in the late summer, we see their webs in our windows and hanging from the eaves of our houses. It helps that our cities offer so many nooks and

A barn spider wraps its prey in a silken blanket. *Photo courtesy of Matt Bertone.*

crannies for our spidery friends. Also, having porch lights on and keeping office buildings well-lit at night brings in the insects, which is a boon for the orb weavers.

Just as the dreamcatcher's nocturnal harvest entirely disappears in the light of day, so to do the webs of barn spiders, who get to work just before dawn carefully disassembling them. They dislodge any uneaten victims and eat their spent silk as they go, recycling it for their next night's hunt.

In the day, they curl themselves up in leaves sewn together with silk to protect them from spider-gobbling creatures like mud daubers, a kind of wasp named for the mud nests they construct. Sometimes in the fall, when the nights grow too cold for a full evening of bug filtering, you might find a fat female barn spider hanging out on the web's hub in the daytime.

A mature male barn spider has a one-track mind and goes looking for a web with a female in it rather sitting in a web all night waiting for bugs to come. If the female's immature, the male will wait by the web while she continues to grow, sometimes for many days. When she's finally mature, he will seize his chance and mate with her. To male barn spiders, getting there first is the best way to ensure they father their mate's offspring.

Asibikaashi worked to protect the Ojibwa children. Barn spider mothers, too, keep their babies safe. When the time comes, they lay up to a thousand eggs in silken egg sacs and meticulously hide them in rolled-up leaves. To you or me, their egg sacs look like one more dead leaf dangling bravely through winter. But inside, her spiderlings cuddle up to each other, surviving winter's chill, protected by their mother's weaving, waiting for spring.

LONG-JAWED ORB WEAVERS (*TETRAGNATHA SP.*)

When stretched out at rest, these spiders resemble sticks. They build orb webs but stretch them horizontally, unlike most orb weavers who spin their webs vertically. Hanging out by streams or other water sources, long-jawed orb weavers may exhibit semisocial behavior, sometimes sharing each other's webs if plenty of food is around. Males have extremely long jaws, which they lock with female jaws, like a pointy kiss, while mating.

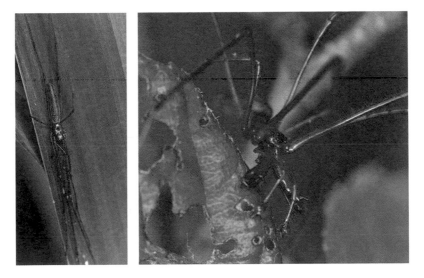

LEFT The long-jawed orb weaver resembles a stick when at rest. Photo courtesy of Matt Bertone.
RIGHT True to its name, the male long-jawed orb weaver shows off his lengthy jaws. *Photo courtesy of Matt Bertone.*

SPINY-BACKED ORB WEAVERS
(*GASTERACANTHA CANCRIFORMIS*)

Like colorful, spiky turtles, these spiders (found in the U.S. South as well as California) are easy to identify. Their tough spines help keep birds away. They can come in different colors, too.

The spiny-backed orb weaver's spines deter hungry birds. *Photo courtesy of Matt Bertone.*

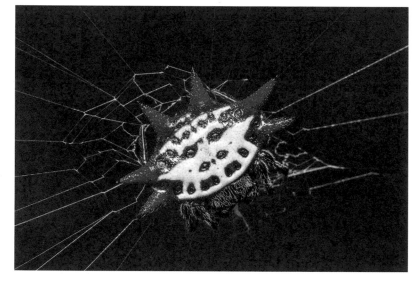

03 BOWL AND DOILY SPIDER

SPECIES NAME: *Frontinella communis*

AKA: bowl and doily spider

BODY SIZE: 0.12–0.16 inches

WHERE THEY LIVE: Bowl and doily spiders are common and widespread across North America but more abundant in eastern parts of the continent. They most often build their complicated webs in low vegetation.

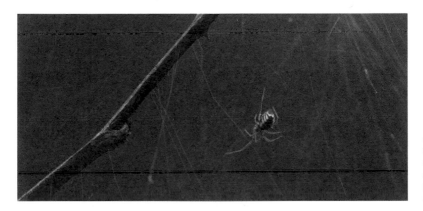

A tiny bowl and doily spider. *Photo courtesy of Daniel Reed.*

Happening on a doily spider web, you might think its habitat is a hot mess. Seemingly fitful jumbles of threads scribble between leaves or fallen pine straw, a ghostly basket of silk thrust almost haphazardly inside. And the spider! That tiny (often less than a quarter inch long), insignificant-looking creature creeping low in the tangle, its black and white patterned-abdomen ticking carefully to the rhythm of its precise brown legs, seems more like a nervous librarian than a remarkable beast. But take a deeper look into that welter web; beneath the bowl you'll find the engineering feat of a spider with a curious, exciting life.

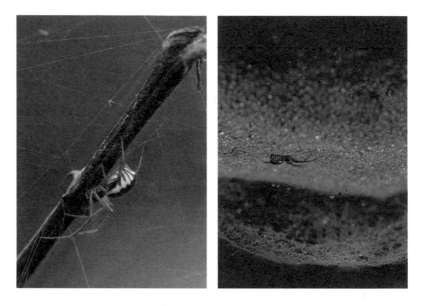

LEFT Much of the bowl and doily spider's web looks like a tangled mess. Photo courtesy of Daniel Reed.
RIGHT The bowl and doily spider waits for a meal on its web's bowl. Below the spider, you can see the doily. *Photo courtesy of Daniel Reed.*

Bowl and doily spider webs aren't sticky. Instead, they're mechanical traps, kind of like pitcher plants or a big catcher's mitt. A disorganized snarl of silk threads tops a smooth, silken bowl. Under the bowl stretches the "doily," a lacy sheet. Tiny flies or other airborne insects, as well as other unlucky passersby, bonk into the tangle and fall to the bowl, where they get stuck. The spiders hang out underneath, eating the ensnared victims through the silk.

They build their webs in forests, preferring openings in the vegetation, near the forest floor. To keep warm, bowl and doily spiders regulate their temperatures by orienting their bodies to align with the sun's rays. When the sun sits high in the sky, they posture their bodies up and down to capture the most heat. Sometimes you may find more than one spider on the web. That's because bowl and doily spider webs serve as love nests for mated pairs. When a male matures in mid-May to late June, he abandons his own web in preparation for moving in with a female. He lets the female know he is the hottest of hot stuffs around by dancing elegantly around the web. He waves his feet at her as he gracefully sways his abdomen around in the air. He shows how smooth and useful he can be by grooming. All of these displays help each female know she has a superfine bowl and doily spider male—and that she probably shouldn't eat him.

Once accepted by a female, a male bowl and doily spider enjoys mating repeatedly and acts like a freeloader, feasting on the female's catch. Male spiders can eat more than 30 percent of a female's food, which sometimes leaves her hungry. The bowl and doily spider eggs that result from these pairings are rarely seen. They can dry out fast, so the females keep them private by hiding them under the soil or under leaves.

With all those fancy-dancing and freeloading men about town, a male bowl and doily spider has plenty of incentive to keep his mate all to himself once he's found her. When other males show up looking for love, the underside of the bowl becomes an ultimate fighting ring where the two spiders have a vigorous showdown. They flex their abdomens to show how big and tough they are. They try to shake their opponents with jerks of the web; they spar with their spider legs; they lock jaws—all while upside down on the underside of the silken bowl. These fights are dangerous. About 90 percent of spars end with serious injury or death to one or both spiders.

Even when they're not worrying about upside-down death matches, life isn't one sweet bowl of flies for bowl and doily spiders. Dewdrop spiders (*Argyodes* spp.) have the nasty lifestyle of sneaking into bowl and doily webs, devouring the homeowner, and taking over the web as their own. In some places, two out of every ten bowl and doily webs have scoundrel dewdrop spiders merrily perched beneath the bowl. Bowl and doily spiders can smell *Argyodes* coming. If they catch a whiff of a villainous dewdrop approaching, they split as quickly as they can.

An invasive cousin, *Linyphia triangularis* moved over to North America from Europe and sometimes shares webs or takes over the more timid bowl and doily spider's home. Time will tell whether *L. triangularis* will completely push out the bowl and doily spider.

Another talented catcher, Yogi Berra, once said, "You can observe a lot just by watching." With bowl and doily spiders, if you take a minute to watch a web, to visually untangle those snarls, to reveal the graceful bowl and enjoy its tidy sheet, to see the cheery, lissome black-and-white bottomed arachnids working on winged animals, you will find a whole marvel-filled world to observe.

PIRATE SPIDERS (*MIMETUS* SPP.)

No spider wants to see members of this spiny-legged swashbuckler genus creeping up to its nests. True to their name, pirate spiders take over other spiders' webs and enjoy smooth sailing, eating the webs' inhabitants, any egg cases lying around, and whatever the poor pirated creatures happened to have captured before meeting their doom. Pirate spiders often attack common house spiders.

This male pirate spider sneaks onto another spider species' web. *Photo courtesy of Sean McCann.*

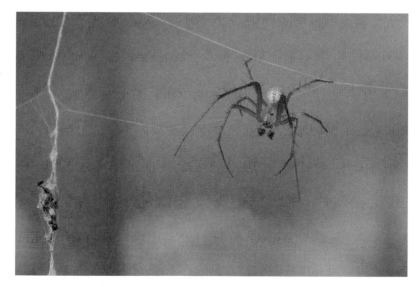

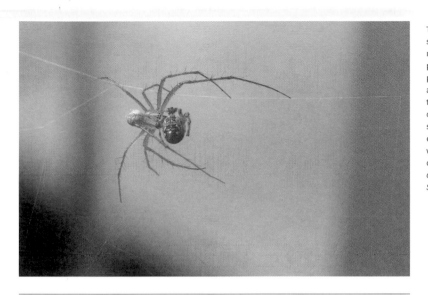

True to his scalawag name, this pirate spider performs a hostile takeover of another species' web, eating the web's previous owner. *Photo courtesy of Sean McCann.*

04 GRASS SPIDER

SPECIES NAME: *Agelenopsis* spp., *Agelenopsis aperta*

AKA: grass spider

BODY SIZE: 0.5–0.75 inches

WHERE THEY LIVE: While *Agelenopsis aperta* are concentrated in the American Southwest, grass spiders can be found across North America north of Mexico. They build their sheet webs in grass or shrubs.

In North America, grass spiders are like little hotel maids, making some of the weirdest beds on the planet, stretching their sheets over earth, stones, low shrubs, and (as their name suggests) grasses.

With an adult body length of one-half to three-quarters of an inch, grass spiders are grayish-brown, with two dark brown stripes along their cephalothorax and two broken, blackish-tan lines running down their abdomen. Two long spinnerets poke out from the

Grass spiders have stripes that run from their cephalothoraxes to their abdomens. *Photo courtesy of Sean McCann.*

ends of their abdomens like Boston terrier tails, wagging out silk from their bottoms.

From July through late autumn, adults wag those spinnerets, furiously spreading silk to form saucer-sized sheets. Each sheet contains a silken tube, a retreat for its inhabitant. This is why these arachnids are sometimes called funnel-web spiders. Unlike many spider webs, the sheets are smooth, not sticky—more like a Velcro-style trampoline than a glue trap. A grass spider waits patiently in her retreat for the gentle pings of insect toes tip-tipping across her sheet. When she feels those vibrations, she zooms out of her hidey-hole and tackles her prey. Grass spiders have long, quick legs built for speed and are some of the fastest spiders in North America. With no sticky silk to keep unsuspecting grass creepers hanging around too long, these spiders must be super quick to capture a good meal.

TOP Grass spiders have long, quick legs. *Photo courtesy of Sean McCann.*

BOTTOM Male grass spiders must be sneaky when approaching females to mate. *Photo courtesy of Sean McCann.*

Grass spiders don't get too attached to their beds once they make them, frequently moving their webs around throughout the season. Ever upwardly mobile, when they spot a higher-quality location, whether it has better food, more food, or more room to capture food, they move right on in.

If you are sneaky you can gently tap a grass spider's web with a leaf or a blade of grass and watch one dart out, hopeful for a meal. But be careful: while grass spiders rarely chomp down on people, their bites can hurt. Unlike most North American spider species, grass spiders are scrappy and territorial. In addition to protecting their webs from intruders, they also defend the area around their webs. When food availability drops, they get the hangries and hang out outside the web more often, hopeful to hop on any little thing scurrying by and spoiling for a spider fight.

Grass spiders' scrappy behavior can get them in trouble. Birds love hopping through fields, shaking webs with pecks and gobbling up the advancing morsels. In some places, peckish birds have been shown to eat more than 80 percent of the local population of grass spiders each week! Some grass spiders manage to avoid the beak by hiding in their silken retreats, where bird beaks can't reach them.

But the weirdest thing about these sheet-weaving spiders is what happens between...er,...on the sheets. Because spiders are predators, and because they often eat members of their own species, mating is a risky business for grass spider Romeos hoping to woo their Juliets. To show they'd make better fathers than afternoon snacks, grass spider males have a complicated courtship.

First, a female will signal that she's in the mood by emitting come-hither chemicals called sex pheromones. A wandering male, getting a whiff of a female's love potion, will approach the web and begin gently flexing it with his excited little feet. The web flexing is a vibrating way of saying, "I'm not just any old grasshopper passing by, lady! I'm the future love of your life! See how deliberately I work this web? It's me! A male grass spider!"

As Romeo flexes the web, he sways his abdomen from side to side in a hypnotic spider love dance. If she seems into it, he releases an airborne chemical, and she passes out cold. She collapses to her web, legs curled up, motionless, deathlike. Seizing the chance, the male runs up, mates with his catatonic lover, and boogies on out of there before she regains consciousness—and possibly an appetite. When she wakes, she stops emitting sex pheromones and prepares to be a mother spider, weaving a thick, silken egg sac into her web.

While the *Agelenopsis aperta* grass spider is most common in the Southwest, dozens of its *Agelenopsis* cousins spread their blankets across North America, making their exciting, strange beds and waiting—waiting for adventure to stumble across their sheets.

05 WIDOW SPIDER

SPECIES NAME: *Latrodectus geometricus, L. mactans, L. bishop, L. variolus*
AKA: black widow spider, brown widow spider, red widow spider, northern black widow spider
BODY SIZE: females are about 0.5 inches, males approximately 0.25 inches
WHERE THEY LIVE: Shy and retiring, widow spiders often construct their cobwebs under rocks, bricks, wood, and other structures. They can occasionally be found indoors and live in urban, suburban, rural, and natural areas. The brown widow (*L. geometricus*) can be found across North America; the southern black widow (*L. mactans*), with its iconic hourglass, can be found primarily in the Southern United States but has been reported across the country and in southern Canada; the red widow (*L. bishop*) is found primarily in Florida.

With smooth, shiny orbs for abdomens and dexterous, fingerlike legs, widow spiders are among North America's most beautiful spiders. Because quite a few (not-so-grieving) widow species hang out

This black widow has captured a green lacewing. *Photo courtesy of Sean McCann.*

all around us, it's helpful to group them by their genus, *Latrodectus*, to examine their most interesting habits.

North American species are dark brown to black in color, with females about a half-inch long (males grow to be about half as large as females). Though we often think of a red hourglass as the sign of a widow, these spiders' abdominal markings can vary exquisitely in shape and in color (from yellow-orange to deep red).

Out of the thirty-three named widow species crawling across the globe, there are three species of black widows in North America: the southern black widow (*L. mactans*) in the southeastern United States up north to Ohio and west to Texas; the western black widow (*L. hesperus*) across western North America; and the northern black widow (*L. variolus*) in the mid-Atlantic United States to southern Ontario. There are also brown widows (*L. geometricus*) and red widows (*L. bishop*).

Usually reclusive, widows build cobwebs in dark, damp, secluded areas such as under rocks, logs, and ledges. They hang upside down in their cobwebs' centers, waiting for insects and other small-to-medium invertebrates to get entangled in their threads. True to their hermit-like tendencies, a widow prefers to have no nonedible visitors to her home. If an intruder bothers her web, the widow drops down like an acorn and plays dead or flings silk at the bothersome visitor as if to say, "Go away!"

We call them widow spiders because legend has it that widow females eat their lovers. Although they sometimes devour their mates (when scientists study them in a lab), they don't do this very often. Their love lives are more complicated and thrilling than their murderous moniker suggests.

A female widow in the mood hangs out in her web, releasing "come and get me" chemicals called pheromones. Males find these chemicals quite attractive and show up at her web ready to go. Males gather much information from these pheromones, including the female's age, how often she's mated in the past, and even how hun-

Widow spider males have striking patterns of their own. *Photo courtesy of Sean McCann.*

gry she might be. The problem is, each female broadcasts her seductive scent as far as she can, and any male around who catches a whiff comes hustling over to hook up with her. The result is a gaggle of love-hungry males scrambling across the web for a shot at mating. Females can pick and choose.

To reduce competition with other males, some sneaky boys creep up to the female's web and reduce the web's size by quickly collecting and wrapping her silk. This diminishes the chemical broadcast, upping the male's chances for mating success. The one who wraps up the web may successfully ward off other males this way, and the female ends up picking Mr. Sneaky in a "I guess you'll do" kind of way. Females can produce many egg sacs each summer, with up to hundreds of eggs in each sac. Widow spiderlings take several months to mature to adulthood, and adults can live for several years.

Because widows hang out in hard-to-reach places, some must be adept at waiting long periods for a meal. By sitting patiently and still, they can last more than three hundred days without eating. When they do eat, they nimbly wrap whatever lands in their webs,

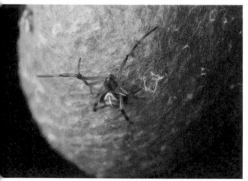

from wayward moths under the eaves to fire ants in cotton fields.

Western widows can be more sociable than their shy cousins and sometimes live in groups of two to eight females in the fall and winter. In their little widow commune, they share webs but not supper. And when it's time to mate, they split up and make their own love nests.

Widows are one of the most feared species in North America because they are known to be venomous to humans. Their genus name *Latrodectus* comes from the venom in their bite, called latrotoxin. Latrotoxin causes humans to have muscle spasms, pain, headache, rigidity, and vomiting—a condition known as latrodectism. Though these symptoms are serious and painful, the bite itself is rarely painful and the effects do not last more than several weeks. However, always seek immediate medical attention if you're bitten by a widow spider. Doctors can help ease the pain with a variety of treatments from muscle relaxants to antivenom.

Though they can bite, widows remain shy spiders and most often run away from danger. One researcher who studies widows tried hard to get individuals to bite by poking and prodding them. Despite bothering them in every way conceivable, the researcher could not get them to bite unless they were squeezed. When they did bite, the spiders more often elected not to inject venom while biting. Even so, always approach widows with a bit of caution. Enjoy their colors and their polished, round abdomens. Watch their dexterous legs fiddle with their cobwebs. Maybe toss them a fly or two. Watch, but maybe don't touch.

TOP
Widow spiders wrap their eggs in a silken sac. *Photo courtesy of Sean McCann.*

BOTTOM
A young widow spider sees the world for the first time. *Photo courtesy of Sean McCann.*

Though often accused of causing terrible medical conditions (such as tissue death and open wounds), these house dwellers rarely bite humans. Restricted to the midwestern and southeastern United States, brown recluses are one of the most feared and incorrectly identified spiders in North America.

Brown recluse spiders are often recognized by the violin pattern on their cephalothorax. *Photo courtesy of Matt Bertone.*

06 CELLAR SPIDER

SPECIES NAME: *Pholcus phalangioides*

AKA: cellar spider

BODY SIZE: 0.24–0.4 inches

WHERE THEY LIVE: The cellar spider builds its webs in nooks and crannies in buildings and homes across North America, even into more northern parts of Canada. In warmer climates, it will also construct its web outdoors.

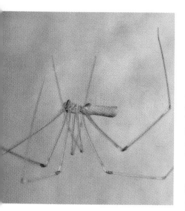

A female cellar spider, hanging upside down in her web. *Photo courtesy of Matt Bertone.*

In the world of your home's basement, cracks, and corners, an unusual kind of killer's at work. But cellar spiders don't really fit the bill of a Hollywood-level assassin. Their tool kit amounts to a wacky "death dance," a silk blanket, and an escape plan that looks like an Olympic figure skater's big finish. Still, these spiders are among the most fearsome spider killers known to spiderkind.

Cellar spiders, the most common North American spiders in homes year-round—north of Mexico—resemble geeky adolescents more than ruthless killers. Eight long, skinny legs spool out from their thin, cylindrical bodies (about a third of an inch long) like an awkward teenager's, looking as if they're always in danger of tripping over their own feet. With pale yellow-brown coloring, they have a gray patch on their cephalothoraxes that sometimes resembles a skull, but even this deathly omen can't lend these spiders a tough-guy appearance.

Like young folks hanging at the mall or bowling alley, cellar spiders sometimes build their webs near other cellar spiders, hanging around upside down in groups. Their webs—tangled, cobwebby clutters haphazardly tossed under eaves, in basements, closets, and

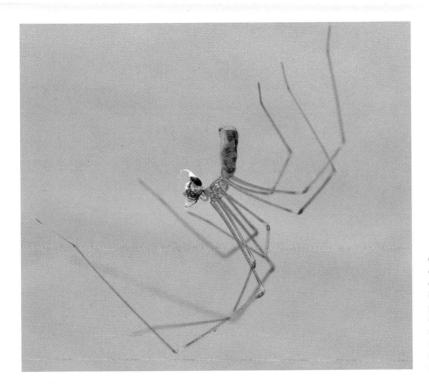

Cellar spiders eat a variety of small invertebrate prey, often enjoying other spiders as a meal. *Photo courtesy of Matt Bertone.*

other low-light areas of human homes—look like a teenager's messy room when compared to the intricate precision of other spider species' webs, like orb weavers or sheet spiders.

Not suited to cold, you'll rarely find a cellar spider enjoying the great outdoors in temperate-to-northern areas (though some people, finding them in their closets, boot them outside!).

Without being sticky, cellar spiders' webs snag household pests like mosquitoes, wood lice, and other insects, so they're good to keep around if you want natural pest control. In addition to making meals of insects, cellar spiders work as methodical spider slayers. They skulk around your home, in search of other spiderwebs. When they find one, they do something called "aggressive mimicry," or a "death dance," where their spindly legs shake their spider prey's web, mimicking the shaking of snagged prey. When a hungry web dweller rushes to the feeding trough, the sneaky cellar spider tosses silk over the unlucky victim, wraps her in a silken blanket, and eats her.

After a meal, the cellar spider lives in her victim's home for a while, feasting on the insects that wander into the web of the dead spider.

When cellar spiders sense danger, they again use their slim form to their advantage. If a predator (or a person) approaches a cellar spider web, the spider quickly drops, keeping his legs in the web, and begins to whirl. Like a figure skater, the spider will swing his body faster and faster in rapid circles, sometimes spinning so fast that the spider seems to disappear into thin air.

Cellar spiders also seem to have an awkward love life. Females mate with many males over the course of their lives. The males do their part by depositing sperm on a little sperm web and sucking it up into a cavity on their palps. They then approach the female and try to hook their sperm web to the proper place on the female. This can be tough for males, and it sometimes takes up to several hundred coupling attempts before the males get the job done.

A female may lay between two and eight egg sacs in her lifetime, each with about twenty to twenty-five eggs. She holds her eggs in her mouth to protect them, forgoing food during their incubation time to keep them safe. She continues caring for them after they hatch, protecting her spiderlings for more than a week as they get acquainted with the world.

Rumors abound that cellar spider venom is toxic to humans. The rumors, however, are false. Cellar spiders rarely bite humans—doing so only if highly provoked—and though their fangs can pierce human skin, the venom only causes a mild burning sensation to those not allergic to it.

Despite their innocuous (even beneficial) relationship to humans, these spindly legged curiosities remain the ultimate spider hatchet men, tapping out their death dance behind our kitchen cabinets.

COMMON HOUSE SPIDER
(PARASTEATODA TEPIDARIORUM)

One of North America's most successful exports, the common house spider can now be found in cobwebbed corners around the globe. Common house spiders make good houseguests and feed on house-dwelling insects. These spiders are close relatives of black widow spiders, but they don't have the dangerous venom of their kissing cousins.

A common corner dweller, the common house spider. *Photo courtesy of Matt Bertone.*

07 FISHING SPIDER

SPECIES NAME: *Dolomedes* spp.

AKA: fishing spider, dock spider

BODY SIZE: females are between 0.6 and 1.02 inches and adult males are between 0.28 and 0.5 inches

WHERE THEY LIVE: Members of this genus can be found across North America.

When I was about eight years old, I was enjoying a nice, warm summer afternoon bobbing around with my friends in the muddy waters of Sleepy Creek, a small lake just outside of town. Always unrealistically hopeful and on the lookout for living creatures to pet and squeeze, I saw what I thought was the shadow of a mouse dart under the rim of the pier.

This fishing spider tests the water for its prey. *Photo courtesy of Sean McCann.*

"I GOT IT!" I shouted, and dog paddled as fast as I could to the pier's edge to catch what I anticipated to be my new, furry best friend. I hauled myself out of the water and bent over the worn boards, peering into their dim undersides, preparing to do a reach-and-grab for something snuggly. Then I saw it. On the pylon, something giant decidedly unsnuggly clung. It seemed to stare back at me with a million creepy eyes.

"EVERYBODY!" I yelled. "It's a MAN-EATING SPIDER! I'll kill it! Y'all swim away!" I bravely swatted off the pylon what I now know is a fishing spider, about the size of a grown man's palm, and I dove into the water to escape its jaws.

What happened next was 100 percent terrifying to someone spider ignorant, as I was at eight years old. When I swatted it into the water, that spider did not drown. It emerged from the water's depths and started *running at us*. Let me be clear: that spider was running at us *on top of the water*.

One of the beautiful things about learning of the ways of our natural world is that understanding the how and why of what we witness can totally reframe our perspective. What had been a frightening experience those many years ago, now fills me, when I look back with an entomologist's eyes, more with amazement than terror, more with questions than dread. That huge creature skated across the water's surface! How did it happen? What was it doing there on the pier anyway? I know now it was minding its own business, the business of a fishing spider.

Among the largest spiders in North America, with bodies measuring more than an inch and legs sometimes spanning as much three and a half inches, fishing spiders provoke fear in many of us who already might be spider squeamish. Fishing spiders are, however, big softies. Well, for the most part, anyway.

Before the days of docks, wharves, and piers, fishing spiders spent their mating season perched on tree trunks and rocks along the edges of ponds, swamps, rivers, and lakes like Sleepy Creek. We have fourteen species of fishing spider in North America north of Mexico, and they're just as happy under your upside down canoe on the shore as they are clinging to a tree trunk. Their camouflage colors—brownish gray mottled with black and light brown markings—look like a stone's surface or tree bark and help them to blend in with their surroundings. This disguise helps them to go unnoticed as they hunt insects, other spiders, and sometimes tadpoles or small fish that they tackle or snatch from the water. They can capture and consume creatures five times their body weight.

With a water repellent body surface, a fishing spider can hide underwater without drowning and chase terrified children… I mean, chase down insect prey across the water's surface. When it dives underwater, water repellent hairs on its exoskeleton trap air close to its body and act as its own personal dive suit/scuba tank combo, giving the spider a silvery sheen and plenty of air to breathe during its subaquatic wanderings. Above water, it spreads its body weight wide across its eight legs and makes tiny boats out of its waterproofed feet, allowing it to tiptoe across the water's surface.

Though relatively large, fishing spiders have no trouble hiding riverside by blending into their surroundings. *Photo courtesy of Sean McCann.*

Fishing spiders can skitter across the water's surface. *Photo courtesy of Sean McCann.*

Despite their fearsome appearance, fishing spiders generally do not act aggressively toward humans but would prefer we leave them alone. Males are monogamous, and females make tender mothers, sometimes going by the moniker "nursery web spider." After mating, a female carries her egg sac in her mouth (or chelicerae) until her young are ready to emerge. Each egg sac can hold up to a thousand eggs. If you spot a mother fishing spider with her capsule of eggs, she will look as if she has a cotton ball tucked beneath her.

When her young are ready to hatch, the fishing spider finds a safe spot, such as a folded leaf or open bark, and weaves a nursery from tangles of silk for her offspring in the space beneath it. She places her egg sac safely in the nursery's center and then waits nearby, guarding her brood as they emerge. Her tiny young enjoy the protection of their nursery for about a week, at which time they molt and strike out in the world on their own, sometimes capturing prey, sometimes becoming the prey themselves to other spiders, ants, wasps, or any of the many spider-hungry creatures crawling about.

As winter approaches, fishing spiders may wander into human

structures searching for warmth. They make startling but good housemates and may devour some pest insect species in our basements before finding quiet places to wait out the winter.

Sometimes I wish I could go back and teach myself things I didn't know when I was a child. Things like: "Take a minute, take a breath, and watch what's happening here before you freak out." And: "Enjoy this encounter! Ask questions!" It would have saved me a few nightmares, some fear of things that never meant me any harm. How great it is to learn things! How nice it is to be filled with wonder rather than fear. So it's okay that I can't teach myself those things back then. I know now. And fishing spiders still wait uncomplainingly on our waters' edges, ready for me to watch and marvel and ask questions.

08 WOLF SPIDER

FAMILY NAME: Lycosidae

AKA: wolf spider

BODY SIZE: with so many species, wolf spiders have a tremendous range in size, from 0.11 inches to 1.77 inches

WHERE THEY LIVE: With more than two hundred wolf spider species in North America north of Mexico, wolf spiders can be found in any habitat across the continent.

Wolf spiders are among the most ubiquitous and conspicuous spiders in North America. Whether a mama wolfie is toting her babies around your garden or a subadult is darting between blades of grass on your lawn, you likely encounter wolf spiders (whether you realize it or not) each day during their peak season, the summer.

Here, a wolf spider has grabbed another spider for a hearty meal. *Photo courtesy of Sean McCann.*

Many researchers around the world have devoted years of their lives peeking into every nook and cranny of wolf spider biology and behavior. Because we know *so much* about wolf spiders, we decided to narrow the field and share with you our top ten favorite facts.

1. They come in every variety. Wolf spiders are extremely diverse, with nearly twenty-four hundred species worldwide and more than two hundred species in North America alone. They all are a little hairy with relatively long legs, but in terms of size, they range from smaller than a dime to bigger than a quarter. They vary in color from gray to brown and often have patterned bodies. They use silk for draglines and egg sac construction, but not for prey capture.

2. They're nothing like wolves. Though they can be a little hairy and they hunt for food, the similarities between these spiders and their lupine namesake ends there. Wolf spiders are loners and don't hunt in packs. And when they "hunt," mostly they sit around waiting for food to show up. When it does, they pounce on it, more tigerlike than wolflike.

3. Wolf spiders live all over the place. With so many different species, they

This wolf spider blends in with its pebbled surroundings. Photo courtesy of Sean McCann.

have adapted to pretty much every condition. Whether they're running up and down beaches, tiptoeing across the Arctic tundra, hunting in farmers' fields, or living it up in the woods, wolf spiders play valuable roles in their diverse environments. They help control populations of other animals, such as small insects, that are lower on the food chain. In addition, animals higher up in the food chain, such as birds, depend on them for food.

4. They have excellent eyesight. Wolf spiders use their eight eyes to hunt and avoid predators. They exhibit eyeshine, which means their eyes glimmer when illuminated at night. Many species are nocturnal and count on their keen vision to help them nab food in the dark.

5. They go with the flow. While many other spider species have fixed life histories (meaning they mate at a certain time of year, produce a certain

Wolf spider females attach their egg cases to their spinnerets for easy portage. *Photo courtesy of Sean McCann.*

number of egg sacs each year, live a certain amount of time, etc.), a wolf spider's life story varies depending on where it lives, not its species. Down South, a single species can have multiple generations in a growing season, while in colder climes it might take two years just to mature. They generally overwinter as immatures and mate during springtime, but this can vary, too.

6. Even without a nest, they're great at nesting. Wolf spiders don't construct webs to provide shelter for their family. Instead, females protect their eggs by carrying their ofttimes bluish egg sacs attached to their abdomens like little Santa Clauses with sacks full of spider babies. With forty or more eggs per sac, that's quite a load of babies to haul around. Some wolf spider mothers of the genus *Pardosa* help speed up their young's development time by sunning their egg sacs on warm days.

7. They make great mothers. After their eggs hatch, the young spiderling wolfies crawl up on their mother's abdomen, where they stay for one to two weeks. All the while, their mama makes sure nobody bullies (or eats) her little babies. Spiderlings don't eat during their free ride, but when they start to get hungry they no longer like to bunk with their brothers and sis-

LEFT It can get crowded on a wolf spider mother's abdomen. *Photo courtesy of Sean McCann.*

RIGHT A wolf spider mother carries her young on her back until they're ready for their first meal. *Photo courtesy of Sean McCann.*

ters and release a little line of silk, from which they float to their new life in the breeze—tiny, daring aeronauts.

8. They're apparently tasty. Many animals enjoy feasting on wolf spiders. In addition to birds, cats, and other larger animals, parasitic wasps specialize in eating wolf spider eggs. When one such wasp finds a wolf spider mama carrying her egg sac, that sneaky bugger will lay her eggs in with the spider eggs. The wasp larvae eat the contents of the egg sac, setting the scene for a sad surprise when, instead of spiderlings, an abundance of well-fed wasps emerge from the mama's sack.

9. They're speedy. Excellent runners, wolf spiders dart across open areas or between fallen leaves. Some species, however, prefer the sedentary life, constructing silk-lined burrows an inch or two beneath the ground where they just hang out, watching the world go by.

10. They're excellent communicators. When it comes to getting their message across, wolf spiders use everything they've got. They use visual cues, like leg waving and dancing, as well as auditory and tactile (well, vibrational) cues to attract mates. Smaller wolf spiders sniff out pheromones of larger spiders and hightail it in the opposite direction to avoid being eaten.

09 STRIPED LYNX SPIDER

SPECIES NAME: *Oxyopes* spp., *Oxyopes salticus*

AKA: striped lynx spider

BODY SIZE: 0.2–0.24 inches for females; 0.16–0.2 inches for males

WHERE THEY LIVE: High abundance in eastern United States from Massachusetts to Florida and Texas; along the Pacific coast to Oregon, and down to Central and South America. They live in agricultural fields (eat crop pests but also pollinators), and in grassy, leafy vegetation.

Throughout North American fields tucked between tall grasses and under petals slink real lynxes, graceful, stealthy hunters on the prowl. Much smaller than the cats who share their name, striped lynx spiders can be just as lissome and just as deadly—to other arthropods, that is.

Like their feline namesakes, lynx spiders are nimble predators. *Photo courtesy of Sean McCann.*

Striped lynx spiders, one of the most common of the eighteen North American lynx spider species, look more like spiky-legged buddies than formidable bug slayers. They vary in color from orange to brown to green and cream, with a darker striped pattern on their cephalothoraxes and spindle-shaped abdomens. Their bodies shimmer with iridescence in the sunlight, giving them the semblance of silver jewels shimmying around field plants. They appear to have two prominent eyes on top of their carapaces, giving them a friendly, puppy-faced look. The spiny hairs protruding from their long legs lend them a bedhead quality. In short, just looking at these quirky creatures, you might find it hard to imagine they could leave a field full of arthropods shaking in their boots. But with striped lynx spiders, looks can be deceiving.

Striped lynx spiders prefer biding their time in agricultural fields. When we plant our crops with only one or two types of plant per field, we humans essentially sow arthropod grocery superstores. In nature, any given species of plant is often mixed in with other plant species and so bugs that like a particular plant species may need to

Lynx spiders are a welcome sight in agricultural fields across North America. *Photo courtesy of Sean McCann.*

search to find the plants they like. As a result, only a limited number of bugs of one type can live in an area. It's like living in a town with a gas station–sized grocery store. In our human-planted superstores, however, tons of insects that like our crops can move into the giant all-you-can-eat buffet of a farm, filled with only their favorite foods. These insects become agricultural pests, gobbling up billions of dollars' worth of food we grow for ourselves each year.

Lynx spiders also have an affinity for our agricultural fields, which makes them helpful buddies for farmers— and for us. They prowl our plants and gobble up all the plant-eating insects they can find, giving us a more bountiful harvest. Though lynx spiderlings spend the field-barren winter months growing up in the grassy margins of agricultural fields after emerging from their eggs in late summer, once crops start to grow in the spring, they move into crops.

Striped lynx spiders can really pack a field. With their shimmering bodies and cat-like pouncing behavior, disturbing striped lynx spiders in a field can sometimes cause an eruption of bouncing silvery creatures. Science tells us that, for every seven lynx spiders in a square meter of field, nearly twelve hundred pest insects are gobbled up each week. With lynx spiders on the job, some farmers saw their pest insect presence drop by more than 50 percent. Field margins that are kept grassy serve as a nursery for striped lynx spiders, which helps to increase the number of eight-legged helpers in our fields.

Striped lynx spiders hunt anything they can get their spiny legs around. Sometimes unlucky pollinators hoping for a nectary treat will fall victim to the spiders' fangs. But, ultimately, they devour far more pests than pollinators. Although they enjoy eating a variety of species at the insect buffet, they prefer to eat whatever insects

Male lynx spiders drum their palps and tap their legs to attract females. *Photo courtesy of Sean McCann.*

they ate as spiderlings. Just as Chris loves his poutine and Eleanor regularly eats collard greens for breakfast, striped lynx spiders remain true to the foods of their childhoods when they can. If we could feed them important pest insects as spiderlings, imagine how we could streamline the striped lynx spiders' hunting in our fields! We could steer them away from pollinators and toward pesky pests like gall gnats.

When it comes to love, a female striped lynx spider is a one-man woman. Striped lynx spiders can mate for up to two hours, but after they finish, the female never shows interest in mating again. Males, however, remain highly interested and trawl the fields for other females who might be interested in his smooth moves. If a male lynx spider detects a female approaching, he'll raise his cephalothorax and drum his palpi on the ground (or the plant leaf, as the case may be) in a "hey, good lookin'! I see you, and you see me, and we're both lynx spiders so why not make the most of it?" kind of way. He drums and pauses. Then, he waves his first two legs up and down as if to say, "Over here! Yes! I'm talkin' to YOU!" The approaching female could decide he's not worth it and reject him, or she could accept him by taking a step toward him and waving her palpae up and down a couple of times in a, "okay, you'll do," fashion. Once he realizes that she's in the mood for love, the male will massage her legs. If she's still into it, he'll mount her and they'll mate. And then he'll be off in pursuit of his next conquest.

Lynx spiders don't live longer than one field season. They dance and mate and hunt the summer long. Females return to the field edges to conceal their egg sacs in the tall grasses. In a few weeks, these soft orbs of eggs push tiny lynx spider babies into the next season, with no knowledge of their father's dance, their mother's hunt, or all the good their kin contributed to the world in seasons before them. The tiny spiders have their own dances ahead of them, their own hunts, and their own good deeds, protecting crops from pests.

FACING, TOP
Woodlouse hunters can often be found scurrying around leaf litter. *Photo courtesy of Sean McCann.*

FACING, BOTTOM
The woodlouse hunter shows its fearsome-looking fangs. *Photo courtesy of Sean McCann.*

WOODLOUSE HUNTER (*DYSDERA CROCATA*)

Introduced from Europe, this nonnative species has made itself at home in the United States, gobbling up rolypolies (also known as wood lice, or pill bugs). Their giant fangs help them grasp this specialized diet.

10 BOLD JUMPER

SPECIES NAME: *Phidippus audax*

AKA: bold jumper

BODY SIZE: 0.5–0.8 inches

WHERE THEY LIVE: Though more common in the eastern part of North America, north of Mexico (from Florida to Ontario), bold jumpers can be found as far west as California.

Evel Knievel was the greatest North American daredevil. Straddling his Harley Davidson motorcycle in his leather jumpsuit, he wowed sold-out crowds with amazing feats of daring. He soared on his motorcycle over rattlesnakes and mountain lions, over rows of cars and vast canyons. And he never used a safety net.

Bold jumpers have large eyes, which they use for stalking prey. *Photo courtesy of Sean McCann.*

Phidippus audax, the bold jumper spider, is the Evel Knievel of the eight-legged world, a seemingly larger-than-life daredevil leaping and flinging itself through fields, prairies, woods, and gardens across North America. Ping-ping-jumpity-jump-jump!

Just as Knievel was a considered by some to be a heartthrob, bold jumpers are robust, good-looking spiders. Adults are stocky, almost like tiny gorillas, and usually three-quarters of an inch long, so it's easy to watch them. Their brown-to-black bodies are hairy with white, triangular spots on their abdomens. Juveniles have orange spots that fade to white as they age. Their chelicerae are stunning, glimmering, iridescent greens and blues.

Bold jumpers have large, charming eyes in the front of their faces, which they use to track small, scurrying invertebrates for daytime meals. They can be picky eaters and are good to have in the garden because while they will regularly gobble down garden pests like bollworms and cucumber beetles, they avoid odor-emitting beneficial bugs like ladybugs and stinkbugs. They've even been known to scavenge on dead insects from time to time but prefer their own fresh-killed bugs, the bold jumper's version of a home-cooked meal.

Bold jumpers don't use webs to capture their prey and are instead active hunters. Their thick, muscular-looking legs make them excellent predators. And boy, can they jump. To ambush prey or escape danger, a bold jumper contracts the muscles in the front part of its body, thereby raising blood pressure and causing its legs to straighten rapidly, which flings the jumper forward. While Evel's longest jump was 144 feet across a fountain outside Caesar's Palace in Las Vegas, Nevada, bold jumpers regularly propel themselves fifty times their body length. That

The impressive male bold jumper shows off his iridescent fangs. *Photo courtesy of Matt Bertone.*

would be like the six-foot tall Knievel flying across three hundred feet from a dead standstill. Take that, Evel!

On particularly dangerous leaps, Knievel would strap a parachute to his back. Bold jumpers use silk instead, attaching a dragline as they move about their neighborhoods. If they make a misstep somewhere up off the ground or need to disappear quickly, they drop and gently fall to safety with the help of their silky strand.

Even a bold jumper can't always escape danger, and they have been known to make a hefty meal for a hungry dragonfly.

Bold jumpers also use silk to build retreats under bark or in other protected areas. They hide out at night in these retreats, as well as in winter as subadults. Females lay about sixty eggs in the protected retreats, keeping them safe from spider-egg lovers like parasites and other hungry creatures.

Evel Knievel didn't have the advantage of a dragline or hydraulics-loaded legs. Still, he held the world record for surviving the most broken bones in one lifetime. Bold jumpers don't have bones to break. They also don't have leather jumpsuits to flash as they soar. But they still lead adventurous lives, tiny daredevils pinging across your lawn.

This beautiful spitting spider waits for her next meal. *Photo courtesy of Matt Bertone.*

SPITTING SPIDERS (*SYTODES THORACICA*)

These ceiling dwellers, common in many homes across southeastern North America, work for their meals by hurling spit at their prey. A mix of glue and silk, their spit bogs down small insects, making them easy to capture. Death loogies are costly for spitting spiders to manufacture, so they save them for special occasions. More often, spitting spiders scavenge on dead arthropods, no slobber-hurling required.

11 ZEBRA JUMPER

SPECIES NAME: *Salticus scenicus*

AKA: zebra jumper

BODY SIZE: 0.25 inches

WHERE IT LIVES: Usually found on or associated with human-built structures—houses, barns, brick walls, schools, town halls. Many times, you can find a zebra jumper in your windowsill. These spiders prefer "dry and hot" habitats, and you can watch them skip along on warm surfaces. Sometimes, zebra jumpers make it to natural habitats like meadows and forests, but they prefer human structures.

Gazing at your windowsill or wandering around your house or school, you may notice a jumping spider hip hopping across the bricks or paint. Upon closer inspection, a dapper little world traveler struts about his business. About the size of a popcorn kernel and dressed in a zebra-striped fur coat, a walrus mustache (or, in spi-

Zebra jumpers seem to watch everything with their large, charismatic eyes. *Photo courtesy of Alex Wild.*

der terms, pedipalps), and with a gait more like a frisky frolic than a creepy crawl, zebra jumpers, *Salticus scenicus*, bear a greater resemblance to dandy gentlemen than to some of their rough and tough cousins. But don't be fooled by their fancy dress and prancing strut: zebra jumpers can punch and pounce with the best of them.

Back in their native Europe, some zebra jumpers gambol around meadows and forests. But the country life just wasn't for them all. Moving into cities and towns, they embraced a more cosmopolitan lifestyle and now happily bask on warm walls and poke around human structures. Having close associations with people can help animals get around. For zebra jumpers, people power, people boats, and people shipping containers helped them hitchhike around the world, and now you can find their big black eyes staring back at you on windowsills across North America.

While we have more than three hundred species of jumping spiders in the United States and Canada, zebra jumpers are among those we can get to know the best. They make frequent appearances in our daily lives because they like to hang out with us. Unlike many invasive species, zebra jumpers make themselves useful houseguests. Gleeful predators—as jumping spiders tend to be—they use their super-sharp eyesight to help them find a good meal. Zebra jumpers pounce and prey on any soft bug smaller than they are and have a penchant for capturing pesky mosquitoes. Because blood-fed mosquitoes fly slower than their lighter, unfed sisters, zebra jumpers make a point of ambushing and gobbling those as a nutritious, but slightly creepy, meal.

TOP
Though dressed in dapper attire, zebra jumpers blend in as they skip along our cement walls. *Photo courtesy of Sean McCann.*

BOTTOM
A zebra jumper chows down. *Photo courtesy of Sean McCann.*

Their keen vision helps them to hunt in the day, another reason we often notice jumpers more than some other spiders. Eight curious eyes, arranged in an unmistakable pattern with two big ones front and center, stare right at us. They can track prey from long distances. And they can spin around quickly to check out any movement, to see if it's something they want to stalk and gobble.

Salticus scenicus, zebra jumpers' Latin name, means "dancing actor," and they certainly do know how to dance. As with most spiders, their leg muscles can flex but aren't very good at stretching out. To get hopping, jumping spiders pump a bunch of fluid to their legs, which builds hydraulic pressure in those twinkle toes. At just the right moment, they release the fluid back into their bodies. The release gives them a bit of pep in their step and helps them to leap sometimes thirty to forty times their body length. That would be like you or me bounding across a football field in one big jump. While this great leap for spiderkind might not be the hottest dance move in town, it helps jumpers ambush prey like pros. Besides, these spiders have more moves than Justin Timberlake.

Zebra jumpers also have a bit of trapeze artist in them, leaving a sticky little trail of silk behind them as they explore the world. If they're frightened or fall, their silk string catches them. With their silk safety harness, they can dangle like acrobats or keep releasing silk as they fall, an aerial dance that makes for a quick and safe way to fly out of harm's way.

The most impressive zebra jumper spider move is the "hey, good lookin'!" dance. When a female spider comes around, jumper males (narrower than females and with giant "spider teeth," or chelicerae) start waving their legs around in an "Ooh! Ooh! Miss Pretty! Look over here, lady! I'm Mr. Hottie, ready for you!" fashion. Then the males prance around like ponies; they wag their abdomens like Bruno Mars's hips; and they tap their toes to a little tune on the ground, all in the hope that their paramour will like what she sees. But trouble brews if another male shows up: a zebra jumper and any of his brother spiders greet each other by spreading apart their giant spider teeth, like arms opened wide. The two males then proceed to engage in what looks like an old-fashioned wrestling competition, where they basically hug each other with their giant teeth and front legs until one gets tired, dies, or gives up. Sometimes they chase each other around waving their fangs and front legs in the air to frighten the other away. The winner? He wins the attention of his lady love.

Zebra jumpers prefer warmer weather and move into your house when it cools outside, preferring to spend their days basking in the sunlight when they're not on the prowl. Look around for them. Watch them dance and spin. Try to get their attention by tapping the ground around them. If you're lucky, you might just glean a fashion tip or two.

Wearing the best ant costumes around, these jumping spiders can hide out safely among groups of ants. An exclusive bunch, ants usually eat or fight whoever tries to hang out with them. Ant-mimicking jumping spiders take advantage of this, blending in with the ant crowd and enjoying the protection of ant mob mentality.

Some ant mimics, like this Synemosyna spider species, can resemble ants so closely, many people mistake them for their ant models. *Photo courtesy of Alex Wild.*

This Peckhamia spider species tries to pass for an ant. *Photo courtesy of Joe Lapp.*

12 CEILING SPIDER

SPECIES NAME: *Cheiracanthium inclusum*

AKA: ceiling spider

BODY SIZE: females: 0.2–0.4 inches; males: 0.16–0.31 inches

WHERE IT LIVES: These spiders share homes and yards with humans anywhere from southern Canada to across the United States.

If you're at home right now—actually, if you're in any kind of building—odds are a ceiling spider is watching you from its perch up in the crown molding of your living room or some other nook in whatever ceiling you currently find yourself under. While we're at

This male ceiling spider shows off his fangs as he crawls along a rock. *Photo courtesy of Sean McCann.*

it, if you're in your yard or in the woods, you're also probably hanging out with a ceiling spider. Never particular over where they pop up their tents, these common creatures operate like a contestant on the *Survivor* TV series gone wild.

We humans sometimes test ourselves against nature in these "survivor" challenges. Some of us sign up to be stranded on a desert island for reality television while others hike mountain trails, and others still simply try having a picnic on the front porch. When thrown against the background of the natural world, we quickly notice the contrast there to the comforts of home.

Ceiling spiders' natural world began outdoors. They still thrive there. Elegant, nimble fellows with smooth-looking creamy pinkish, orangish, brownish, or grayish bodies, their shiny, dark brown chelicerae and eye area augment their darker, orange-tinted cephalothoraxes.

They have a darker but still creamy dagger marking cleaving the top of their abdomens, causing them to look as if they're lithely dragging one big heart behind their heads. They look almost completely naked, save for luxurious black hairs tipping their toes. Their dark feet blend into leaf litter and make it appear almost as if the spiders are levitating as they scurry over the ground.

Outdoors, where ceilings do not exist, these spiders spend their days rolled up in leaves or litter, their own homemade sleeping bags. At night, they emerge to hunt flies or other small arthropods. No flies? No problem. Unlike most strictly carnivorous spiders, ceiling spiders sometimes dip their fangs into flower cups, enjoying sweet, floral nectar. Omnivory increases this species' ability to grab a meal or a snack wherever it can, helping it thrive where other spider species go hungry.

Although quite at home under the stars, ceiling spiders have experimented with their own version of *Survivorspider*, testing their survival skills in the wilderness of our homes. It turns out that for ceiling spiders, even without syrupy flowers to drink or leaves

Ceiling spiders have black "toes." *Photo courtesy of Sean McCann.*

in which to wrap themselves, camping in our homes is more like "glamping," camping in the lap of luxury.

Instead of leaf-made sleeping bags, they build silken retreats for themselves in our crown molding. During the day, they lounge extravagantly in their silk tunnels, waiting for evening.

As the sun goes down, ceiling spiders go on the hunt. Many of us would like to believe that we live alone, perhaps with our loved ones and maybe a pet or two. We think of our homes as sterile environments that rarely admit any other critters. The truth is that most of us have more than a hundred different arthropod species sharing our home. We live amid a jungle of creatures, from curious cockroaches to fierce, tiny wasps that specialize in having babies that gobble up cockroach eggs. Unusual, primitive-looking insects trundle across our windowsills, eating leftover pollen grains that blew in last spring. Tiny beetles gambol around between our carpet fibers, and long-legged crickets bonk around in our basements.

Each of our houses is its own ecosystem, with herbivores, predators, parasites, and all other links in the food web that depend on each other to survive. Though ceiling spiders have neither nectarous flowers indoors nor the abundance of wild bugs they're used to outdoors, they still have plenty to eat. At night, they prowl for fruit flies, houseflies, or anything else they can find, taking a meal while we rest. By eating insects we consider pests, ceiling spiders help us have a cleaner home while at the same time, of course, indicating that pests are present there.

You can sometimes clue in to a ceiling spider's diet by checking out its color. For example, if he eats a bunch of houseflies, he has a grayish appearance. If instead he snags all those red-eyed fruit flies that appeared from that banana you left out a little too long last week, he has a pinkish hue. Next time you find a ceiling spider in your midst, see if you can guess what it's been eating. Then thank it for a job well done.

Sometimes its ability to live with humans has gotten us into trouble (the spiders, however, seem to come out just fine). A few years ago, a car manufacturer recalled tens of thousands of cars, reporting that the eight-legged adventurers spun their welcome mats in vent lines, potentially causing engine damage. To prevent any more unwelcome spidery visitors, the company designed a special spider stopper to plug the vent.

Ceiling spiders have also been implicated in spider bites. However, the odds of a ceiling spider biting you are extremely rare. While ceiling spiders inhabit millions of homes across North America, only a few people report being bitten by one. If a ceiling spider bites you, it may hurt like a bee sting, though the venom is very different. For people in good health, the redness and pain won't last and there won't be any long-term effects.

Even if you run the extremely slim risk of a ceiling spider's nip, they're worth keeping around the eaves for the fly control. If you decide you'd rather not have a spider napping in your crown molding, consider tossing your ceiling spider outside, provided it's not winter in Montreal. Their glamping days behind them, ceiling spiders are one of those house-dwelling species who can dust themselves off and make themselves right at home back in the great outdoors. As for us, we'll stick to the conveniences of home, sweet home.

13 GOLDENROD SPIDER

SPECIES NAME: *Misumena vatia*

AKA: crab spider, goldenrod crab spider, flower crab spider

BODY SIZE: up to 0.38 inches for females; up to 0.13 inches for males

WHERE IT LIVES: Temperate North America, along the boundaries of grass and shrubs, open locations, and generally around yellow or white flowers, especially goldenrod and daisies. They often prefer to lay and guard their eggs on milkweed plants.

Some of us might think pollinators have life pretty easy. Flitting from flower to flower, sucking up sweet stuff, rolling around in pollen, fluttering around in soft petals. Some, like butterflies, appear resplendent with giant, colorful wings; others, like bees, seem nearly invincible with those stingers. Such a lovely life! What many of us

This unlucky fly thought it was landing on a flower but, instead, became a flower crab spider's supper. *Photo courtesy of Sean McCann.*

don't know is that sprawled out on those velvety petals, hiding in plain sight, a deadly assassin patiently waits. Powerful venom tucked in her maw can render even stingers useless. Raptorial spines that are folded into her grasp eviscerate beautiful batting wings. Crab spiders, among the most common spiders in North America, act like hired guns, spies infiltrating flowers and specializing in nabbing pollinators that land for a meal.

About the size of a plain M&M (a little smaller for males), crab spiders do indeed resemble miniature crabs. Thick, crablike legs hang from their globose, compact, crablike bodies. They can scuttle about in all directions like crabs, running front, back, and sideways. Female goldenrod crab spiders can be white or yellow with two pink bands running along their abdominal sides. Rarely seen males are much thinner and often have darker cephalothoraxes but maintain a lighter abdomen. Like any good superspy assassin, crab spiders are masters of disguise. We have nearly seventy crab spider species in North America north of Mexico, and all demonstrate the incredible talent of hiding. Some crab spider species have patterns on their

By shifting pigment in their cells, crab spiders change colors to blend in perfectly with their flower-head homes. *Photos courtesy of Sean McCann.*

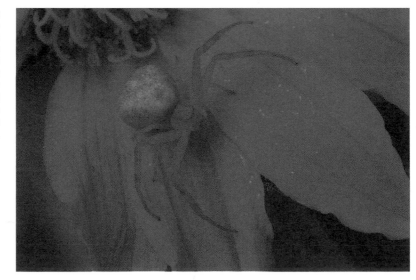

bodies that resemble their habitats. Others squish flat to fit perfectly in plant or bark cracks. Goldenrod crab spiders have the best disguise: they change colors to match their backgrounds.

When crab spiders find themselves on plants where their body color stands out, they scurry around until they find a good spot that matches their coloration. Sometimes, though, running to the right color just doesn't work, so crab spiders put on their best disguises by changing their body color from white to yellow or anything in between to blend in. They do this by shifting pigments in their bodies until they reach the right hue. Unlike chameleons, who can change colors in minutes, or super spies, who can don their fake mustaches and overcoats in the blink of an eye, crab spiders take about three days to color match their bodies to their environments.

Once they put on the perfect disguise, goldenrod crab spiders sit on flower petals and wait for hapless pollinators to arrive. They wait. And they wait. And they wait. Sometimes, crab spiders can wait several days without a meal. When unlucky pollinators arrive, WHAM! Crab spiders jump on them like a monkey on a doughnut. They use special spines in their raptorial forelegs to ensnare their prey. Although they have weak chelicerae, these spiders have very strong venom. They use their toxins along with their spiny legs to take down pollinators much larger than they are, like bees.

Unlike many solo-operating 007-type spies, crab spiders have active family lives. Because female crab spiders spend all their days undercover on flower heads, males must wander around to find a lover. Imagine these males, wandering from flower to flower having hopes they might one day wander up to the right flower and find a big, fat female sitting up there.

For a male crab spider, stumbling on a female crab spider sitting on her flower is like winning the lottery. With all the flowers out there, the odds are pretty low that he'll climb a stalk with a female hiding on top. If he hits the jackpot, he doesn't want to lose his winnings to another male. He spends the rest of his days guarding his girl from other love-seeking males, trying to battle any fellow attempting to climb his flower stem.

In the interest of keeping her offspring undercover, female crab

Crab spiders often snag pollinators. This one snagged a bee. *Photo courtesy of Sean McCann.*

spiders relocate from their flower heads to a different, leafy plants to provide shelter for her cocoon-like egg sac. She affixes her eggs to the underside of a leaf with silk and guards her young until they hatch. Crab spiderlings scurry out into the world and try to avoid being eaten by spider eaters such as other spiders, wasps, and ants or being smashed by frightened humans bringing yard-picked bouquets into their kitchens. The spiderlings capture larger and larger prey as they grow, using the same ambush techniques of their parents. Those that survive will one day become beautiful assassins and terrors in flower heads, like their mothers, or optimistic wanderers like their fathers.

This black spider (and its Western cousin, *H. hesperolus*) looks like it's always ready to receive a confession, with its distinctive, white chevron markings that resemble a parson's collar. Shy and speedy, parson's spiders quickly zip between hiding places, both indoors and out.

This parson spider is ready for confession. *Photo courtesy of Sean McCann.*

14 BLACK PURSE-WEB SPIDER

SPECIES NAME: *Sphodros niger*

AKA: black purse-web spider

BODY SIZE: 0.4 inches

WHERE IT LIVES: Though its cryptic habits make it difficult to pin down its distribution, this spider is evident in the eastern United States, possibly to southern Canada.

We want to let you in on a big secret: Whenever you take a walk through the woods in eastern North America, you may be stepping right over a mysterious, fearsome-looking spider, one so furtive that we know very little about it. Black purse-web spiders may not make the most-common list, but their existence tells a story about the rich

Sphodros are mygalomorphs, which means they're more closely related to tarantulas than most other North American spider species. *Photo courtesy of Matt Bertone.*

natural world hidden from us, waiting right there beneath our footprints.

Unlike most spider species, and unlike more than 95 percent of North American spider species north of Mexico, black purse-web spiders are mygalomorphs, a "primitive" type of spider. To scientists, primitive doesn't mean this spider is still making cave paintings or hanging out with dinosaurs. Instead, it means that it shares physical characteristics with its early spider ancestors. For example, black purse-web spiders, like all mygalomorphs, have parallel fangs, kind of like your pointer and middle finger, which close in an up-and-down, vampire-like motion. Araneomorphs (the majority of North American spiders, and all other species discussed in this book) have fangs that close in a pinching motion, like your pointer fingertip meeting your thumb tip.

Other mygalomorphs include tarantulas and trapdoor spiders. Black purse-web spiders share the chunky, muscular appearance of these mygalomorphs, with broad, somewhat flat cephalothoraxes

Black purse-web spiders have long fangs like this Sphodros species. *Photo courtesy of Matt Bertone.*

and thick legs that look almost like human fingers spreading from their middles. True to their moniker, they're black all over, with matte-black cephalothoraxes and abdomens and shiny black legs.

Female bodies are an inch long, which is twice as long as males. Their long fangs look like frightening sickles tucked under their bodies, but don't be afraid. Odds are you won't bump into one of these creatures anytime soon, and, if you did, they're more likely to run away than attack. These spiders are so secretive and so rare that even though humans first described a male, black purse-web spider in the 1840s, females weren't described until nearly 140 years later, in 1980.

They may be more common than we think, but they're tough to spot because of their shy nature and secretive lifestyles. Instead of the showy webs boasted by some araneomorphs, such as barn spiders, black purse-web spiders build silken tubes between four and six inches long (though purse-web spider tubes have been found that are up to two feet long!) extending underground from the hidden, organic layer below pine needles in pine forests. Sometimes these tubes may extend as high as a tree base. A black purse-web camouflages any exposed part of his or her tube by covering it with debris. This is another reason they are so hard to find!

Black purse-web spiders sit in their tubes like mythical monsters hiding in caves, waiting for some unlucky traveler to cross their paths. When prey such as wandering millipedes (a black purse-web spider favorite) cross over the tube, the black purse-web spider strikes through the tube wall with her long fangs, impaling her meal through the silk. She then cuts a small slit to drag her quarry through to the darkness of her home. Quickly, she patches up the silk so no trace of the violence remains and other unsuspecting meals won't be tipped off to what lurks beneath the leaves.

When she finishes her snack, the black purse-web spider may toss her prey from the tube. Though often, opened tubes reveal cast skins and the spent remains of spider meals stuffed aside.

Black purse-web spiderlings hatch in the fall. They remain hunkered down with their mothers for months until the following spring, when they strike out to build their own secret tunnels.

This species is so cryptic that the Committee on the Status of Endangered Wildlife in Canada lists it as a potential species of concern, but the truth is we don't know how rare it actually is. It's possible that black purse-web spiders are around and we just aren't seeing them. What other pieces of life aren't we seeing? What exciting stories loiter undiscovered for each of us beneath the leaves? For those of us who think humans have already revealed every living thing around us, black purse-web spiders are a good sign that we've only scratched the surface.

SLENDER CRAB SPIDERS (*TIBELLUS OBLONGUS*)

Sprinkled in shrubs, vegetation, and gardens across North America, these common spiders look like pale, striped letter *X*s as they wait for prey with two leg pairs stretched out front and two leg pairs stretched behind. Their longitudinal stripes help them hide in the weeds.

Sticklike crab spiders make themselves almost invisible. *Photo courtesy of Sean McCann.*

FREQUENTLY ASKED QUESTIONS

Chris answers the ten questions most asked of arachnologists.

Is a daddy longlegs a spider?

Like spiders, daddy longlegs are arachnids, have eight legs, and are common in forests, fields, caves, and other dark places. However, daddy longlegs are not spiders. Often confused with spiders, daddy longlegs are classified in a different scientific order (Opiliones) than spiders and have many different physical attributes. Opiliones members (daddy longlegs, also known as harvestpersons or shepherd spiders) don't have a spider's narrow waist or pedicel. They look like their bodies are made up of one big section (compared to spiders' two sections). Legend has it that daddy longlegs are venomous but, unlike spiders, these creatures have no venom glands. And unlike their spider cousins, these critters don't make silk and prefer to scavenge for a meal instead of hunting live prey.

There is a daddy long-legs spider (also called a cellar spider) that at first glance resembles the true daddy longlegs. If you look carefully, though, you will find that cellar spiders (see page 34) have two body segments and hang out in webs.

Is a tick a spider?

Ticks are not spiders. Although arachnids (like spiders, scorpions, and daddy longlegs), ticks belong to the scientific order Acari (which also includes mites). They differ from spiders in many important ways, including that they appear to have only one body segment, are generally quite small, and often have a very hard exoskeleton.

Can all spiders bite? Are all spiders venomous? If they bite me, will I get a bacterial infection?

While it's true that all spiders are venomous, not all spiders can bite humans (remember that many species are very, very tiny!). Even those species that can bite humans rarely do. Spider venom is mostly suited for their insect prey and not generally strong enough to affect humans.

In North America, there are only two groups of spiders that may be "medically important"—the brown recluse (see page 33) and the widows (see page 29)—which means these two spider types have the potential to cause serious harm to humans. Even with these groups, serious medical complications are uncommon. In many cases of "harm," a diagnosis of a bite from these spiders doesn't even coincide with the range of the species. It's easy to blame a spider, but an accurate verification of a spider bite really requires capturing the culprit and identifying it.

Research has shown that misdiagnosis of bites or other wounds as having been caused by spiders has led to unwarranted fear of spider bites. Because of this, medical practitioners should look to other more plausible explanations, perhaps bedbugs, fleas, or bacterial infections.

You often hear that a spider bite might result in a bacterial infection because a spider's fangs are "dirty." However, a recent scientific study found that this is not likely. That being said, spiders do sometimes move around the globe on our bananas or hidden in grapes. Thankfully, arachnologists have found that very few of these spiders are dangerous to humans. To avoid being surprised by globetrotting arachnids, it's a good idea to keep an eye out for spiders when selecting your fruits and veggies.

Ultimately, you just don't need to wander around being worried about spider bites. Most spider species are small, walk unnoticed among us, flee to dark hiding places, and are far more wary of us than we need to be of them.

How do I become an arachnologist?

You have already made one important step forward by reading this book. Now that you have met a few spiders, you can graduate to more detailed field guides, books, and online resources.

A couple of excellent spider books are Richard A. Bradley's *Common Spiders of North America*, and Rainer Foelix's *Biology of Spiders*.

If you want to learn species, try out *Spiders of North America: An Identification Manual* by Paula Cushing and Darrell Ubick. Using that book requires you to seek out a microscope, so you should ask in your local university or college to see if you can find a lab that will let you use their space and tools to work.

For more advanced training, you may wish to embrace the field of entomology (the study of insects) as many of the same folks who

Chris Buddle scours the landscape for eight-legged wonders. *Photo courtesy of Crystal Ernst.*

love insects can help you travel into the world of spiders: many colleges and universities have classes about insects, and sometimes you can get lucky and learn about spiders too.

You can also visit the website for the American Arachnological Society or the International Society of Arachnology and access many resources, from photographs to scientific publications and announcements about upcoming scientific conferences.

You can also become an arachnologist in your own backyard. You don't need many tools to start! Get yourself some empty pill bottles, tweezers or forceps, and perhaps a hand lens (a small, high-powered magnifying glass) and start your own spider collection.

You also need a sturdy notebook for collecting data and for making behavioral observations. If you have a smartphone, you may wish to collect data digitally and perhaps connect with a broader community of people using apps such as iNaturalist.

You can collect spiders by picking them up, either with or without tweezers; you can also shake a shrub or sapling over top of a white sheet and see what falls off. You may eventually purchase a sweep net from an entomological supply company like BioQuip and gently sweep grassy areas or hedges for spiders. It's always productive to go spidering at night, as wolf spiders have eyeshine when you shine a flashlight on them—it's an easy way to track down nocturnal hunters. Check around porch lights too: those are terrific areas for high concentrations of insects, so you can bet that some big orb weavers will set up shop nearby.

You can do some amazing and easy fieldwork with spiders: observe your local spiders carefully, both in their natural habitats and after you collect them. You can make observations about spiders by merely sitting down for a while and watching them: perhaps you can find an ant-mimic jumping spider by keeping a close eye on an anthill. Maybe you can catch a glimpse of an orb-web spider capturing and eating its insect prey: What kind of prey does it collect? Are the prey smaller or larger than the spider? Is the spider active at

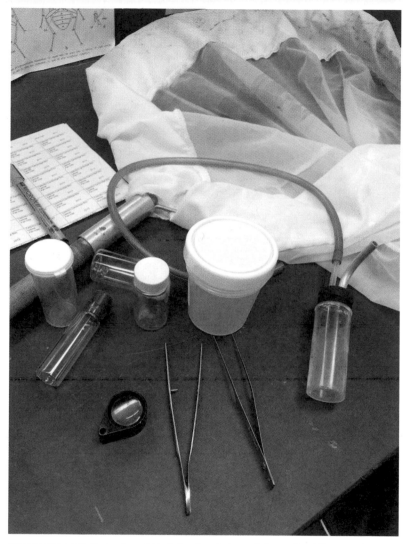

Common spider-collecting equipment (clockwise, from the bottom): forceps (for picking up spiders), a hand lens (for looking at them up close), some vials (various sizes, for holding specimens), a pen and labels (for writing down where the spider was found), a sweep net (great for collecting all kinds of spiders and insects), and an aspirator (a device used to suck a spider into a collection vial; also sometimes called a "pooter"). *Photo courtesy of Chris Buddle.*

Chris Buddle gets up close and personal with the landscape to find spiders. *Photo courtesy of Johann Wagner.*

night or during the daytime? You can do a lot of great arachnology in a city center, too, since many spiders are fond of the concrete jungle, tucked away behind neon lights or lurking in stairwells. Do you see different species at different times of the year? Can you find different types of spiders in different parts of your local skateboard park? How many grass spider webs can you find on the hedge outside your uncle's townhouse, and does the number vary from year to year?

If you want to keep your spiders for later study, arachnologists typically store dead specimens in some kind of alcohol. In the beginning, you can purchase rubbing alcohol from your local pharmacy. You will probably need to track down some small glass vials that have a lid with a good seal. Keep careful notes about your specimens and always write a label for your spider so you can tell where you collected it from and when. Also make a small label that includes your guess about the kind of spiders you collected. This can be general at the start (e.g., "jumper!") but as you learn more about spiders, you may be able to identify it to the species level.

Many towns and cities also have naturalists clubs: think about joining these, and hang out with other people interested in nature. Go on local hikes with these groups and soak up as much about

LEFT A scientific collection of spiders. Specimens are kept in ethanol and stored in individual glass vials. These particular spiders were collected from the Arctic, as part of Chris Buddle's research program. *Photo courtesy of Chris Buddle.*

RIGHT In a research collection of spiders, all species from the same family are usually housed together. This drawer contains jumping spiders of the family Salticidae. *Photo courtesy of Chris Buddle.*

nature as you can. Naturalists often have broad interests, and chances are good that you will find a friend who is also keen to learn more about spiders, or, better yet, you may discover that a fellow arachnologist is already a member of the club.

Don't be shy about calling up your local college or university to ask if anyone in the biology or entomology department has an interest in spiders.

You can also use social media to learn about spiders: check out Twitter or Facebook for arachnologists. The community is always happy to share their passion with enthusiastic arachnophiles.

How do I keep spiders as a pet?

Spiders make interesting pets, but like all pets they require attention and care. Start by making a home for your spiders: this can be a glass

Spiders, like these egg-carrying wolf spider females, can be carefully captured in vials. *Photo courtesy of Chris Buddle.*

jar, or something fancier like a glass or plastic terrarium (these can be purchased at a local pet store). Put some potting soil in the bottom. Adding a stick or two is a good idea so your spider has a place to crawl. Sticks are especially important if you are making a home for a spider that builds a web because these spiders need some kind of structure on which to anchor their webs. Spiders also need fluids, and an easy way to provide humidity and water is to stick some cotton into a thin pill bottle (without the lid), wet it, and place it in the bottom of your spider home. Check the cotton regularly and make sure it doesn't dry out.

The biggest issue with keeping spiders as pets is that most spiders need live food to survive.

If it's spring, summer, or fall, you can feed your spider by finding other insects in your yard, or even in your home. Spiders generally like soft, live prey. Although many spiders can tackle prey much larger than themselves, it's generally a good idea to find smaller food items such as houseflies, crickets, or caterpillars.

When the colder season arrives it might get tougher to find live food on your own. This means you may need to take a trip to your local pet store to see about purchasing some crickets or other live prey.

Sometimes it's wise to just keep your spider for a few days to observe it and become familiar with its habits. After a bit, you may consider placing it back in the wild. If you decide to keep it for a longer time, don't get too attached to your spider: most species don't tend to thrive for too long in captivity.

An exception to that rule is tarantulas: these big, hairy spiders can be purchased at your local pet store but make sure to do your homework. Do some online research and make sure the store where you purchase your tarantula doesn't stock rare or endangered species. If you manage to get a female (you can't tell if the tarantula is immature!), your pet tarantula can live for many, many years. They will molt periodically, even as adults, and some species are quite friendly and can be handled.

Although your spider won't need walks every day, you do need to keep its terrarium clean, to maintain moisture, and to provide live food. As with all pets, having a tarantula is a commitment.

Spiders, such as tarantulas, can make great indoor companions. Harriet, a twenty-year-old Chilean rose hair tarantula, has been living in a terrarium in Chris Buddle's lab for about fourteen years. She likes to eat crickets, which can be purchased at a local pet food store. *Photo courtesy of Chris Buddle.*

What eats spiders?

Although we think of spiders as being rough-and-tumble predators, many animals prey on spiders. Birds do, and in many parts of the world, overwintering birds would have a tough time getting through the cold season if it weren't for spiders.

Many species of specialized wasps also feed on spiders, and the stories are certainly horrific enough to terrify young spiderlings from around the world. Some wasps seek out spiders, paralyze them, and carry spiders back to their nests. These paralyzed spiders, in their half-dead zombie state, make perfect food for the wasps' young. Yummy.

Some wasps also specialize on the eggs of spiders: in this case the mother wasp lays its eggs inside the egg cases of wolf spiders (see page 46). The young wasps do a pretty nice job of feasting on all the eggs in the egg sac, and the poor mother spider is left without any babies at all. Instead of seeing all the spiderlings emerge from an egg case, out pops more spider egg–hungry wasps.

Some flies also specialize in preying on spiders. The maggots of one group, called small-headed flies, grab onto a spider as it walks by. The maggots burrow their way into the spider's abdomen, making a home inside the spider. Eventually the maggots pupate into adult flies and emerge from the spider: not so good for the spider, but an interesting life-history strategy for the flies!

What do spiders do in the winter?

Spiders are quite hearty and are even found in parts of the world covered by snow for much of the year. This includes the most northern parts of the world. However, even in more temperate regions, spiders deal with freezing temperatures during some months.

Many spiders do their best to wander to warmer places as temperatures get cooler during the autumn months. Sometimes this includes our basements. That's why it's not uncommon to find a lot of spiders in our houses in the fall. Spiders also find warmer habitats in wild areas, and this often means nestling down in the leaf litter or under bark.

It is generally thought that most spiders are "freeze avoidant," which means they cannot tolerate being frozen and will die if temperatures drop below zero. The adaptation for dealing with cold temperatures therefore involves a fascinating process called supercooling. This means that spiders can produce a type of antifreeze in their bodies, which enables them to stay unfrozen when temperatures dip below freezing: it's a neat trick shared among many arthropods as a way to adapt to winter.

Many spiders remain active in the winter, sometimes popping out on top of snow when an unusually warm day appears in February. Snow also provides an excellent insulation for many arthropods, and some species of spiders thrive in the small space between ground and snow.

Other species, however, can't survive the winter very well and instead let their protective egg cases shield precious spider eggs during the long cold days of winter. The writing spider, for example (see page 11), employs his strategy, and although females die before winter hits, their egg sacs overwinter without difficulty.

What do I do with spiders if I find them in my house?

If you care about spiders, the best thing to do is let any spiders you find hang out in your house. After all, these animals help to control other insects in your house. But, to be fair, the spider poop can make a bit of a mess, and if you are an arachnophobe, or live with one, you may need to deal with the situation.

If you don't want to handle your spider but need to move it, do the old "cup over paper" trick, place a cup over the spider and gently slide it over a piece of paper so the spider is now captured. But what to do now...

It gets a little tricky: people generally have a belief that an indoor spider is really just an outdoor spider that is somehow trapped inside. They feel they're helping the spiders to "free" them outdoors. But remember, many spiders found in our homes are actually in their preferred habitat already (e.g., the cellar spider—see page 34), and sometimes it's much colder (or hotter) outside than inside. So even though releasing your spider back into the wild may make you feel better, you are probably actually killing your spider by releasing it into the wild outdoors.

Perhaps the best thing to do is move your house-loving spider to a part of your house where it can live peacefully and without bothering you: perhaps your basement or a sunroom with lots of houseplants or a closet. Give your spider a room of its own.

How come spiders don't stick to their own webs?

Spider webs come in all shapes and sizes, and many webs are not actually sticky at all! For example, grass spiders (see page 26) or bowl and doily spiders (see page 21) use webs to capture prey but don't need sticky fibers to trap their meal.

However, some spider webs are super sticky, such as classic orb-weaver webs (e.g., see page 11). Even though their webs are adhesive, the spiders themselves are able to move about with ease. What's the trick?

Well, the trick isn't all that tricky: it's practical. Many spider threads aren't sticky for their entire length and instead have small droplets of "glue" along their length. In the case of the classic orb webs, for instance, the center "hub" and spokes of their webs are not sticky whereas the spiral strands are. The spiders know where to walk to avoid their own gluey spots, and instead of rambling through their webs like a bull in a china shop, they only touch their webs with their dainty tiptoes. They show great agility and control and in this way avoid sticky parts of their webs.

Spiders also have terrific hygiene: if you watch spiders long enough, you will notice they will often run their legs through their mouths, as a way to keep their feet and legs meticulously clean, thereby making them less prone to getting caught in their own webs.

I've heard that when you sleep, spiders crawl into your mouth and you probably eat thousands of spiders throughout your lifetime. Is this true?

No.

ACKNOWLEDGMENTS

A sincere thanks to all the arachnologists who inspired us and helped with the project. They include Robb Bennett, Rich Bradley, Paula Cushing, Joe Lapp, Sean McCann, Catherine Scott, Rick Vetter, and a large and engaged social media community. Thank you to the talented and generous photographers who contributed to this project, especially Matt Bertone and Sean McCann; your work inspires so many people, including us. Thanks also to Munib Khanyari for doing some of the background research for us. We are appreciative to Robb Dunn and the team (including Holly Menniger and Neil McCoy) at North Carolina State University. Finally, the Buddle family and Spicer Rice family supported us and our crazy spider obsessions— huge thanks to Greg, Pete, Thompson, Shannon P., Becky, Evan, Emma, Peter, and Robin Anders.

GLOSSARY

aciniform: a spider's silk gland, which produces the type of silk that wraps and secures prey, as well as forms the stabilimenta and male spider sperm webs.

abdomen: a spider's "rear end." The end segment (i.e., second of its two parts) of a spider's body, behind its cephalothorax.

aggregate: the spider gland that produces a sticky silk glue.

ampullate: the silk gland that produces nonsticky dragline silk or temporary scaffolding silk.

arachnid: any animal in the scientific class Arachnida. All arachnids have eight legs, but some arachnids have appendages that look like extra legs. Members of the same scientific class are like distant cousins, sharing more physical and behavioral characteristics than they do with nonrelatives, but not as many as they share with close cousins, those in the same scientific order; extended family, those of the same scientific family; or immediate family, those of the same scientific genus. Spiders are arachnids, as are ticks, whip scorpions, daddy longlegs, scorpions, pseudoscorpions, and sun spiders.

arachnologist: a person who studies arachnids, including spiders.

arachnology: the study of arachnids, including spiders, ticks, whip scorpions, daddy longlegs, scorpions, pseudoscorpions, and sun spiders.

Araneae: the scientific order to which spiders belong. Members of the order Araneae have eight legs and venom-injecting fangs on their chelicerae.

araneologist: someone who studies spiders. In truth, many people use "arachnologist" to mean "araneologist," essentially because the latter is a seldom-used word. However, if we want to be pedantic about things, if you are studying just spiders, the proper term is "araneologist"!

araneomorph: any spider in the scientific classification of the infraorder Araneomorphae. Most spiders (more than 95 percent of North American species north of Mexico) are araneomorphs. The remaining spiders are classified in the scientific infraorder Mygalomorphae. Araneomorphs have fangs that bite in a pinching action, rather than running parallel to each other and biting in an up-and-down fashion, as in mygalomorphs.

book lungs: organs used by spiders for breathing, located beneath the abdomen.

bridge line: the silk line in a spider web that bridges the web's upper points.

carapace: the exoskeleton that covers the spider's cephalothorax on its dorsal (or back) surface.

cephalothorax: the front end of a spider's two main body segments. It includes the fused head and thorax segments and possesses the legs, eyes, mouthparts, and all legs.

chelicera (pl., chelicerae): a spider's main mouthparts, consisting of the venom-injecting fangs attached to the cheliceral base.

dragline: a nonsticky strand of silk produced by the ampullate glands used for lowering a spider to safety or around its web.

embolus: a male spider's external reproductive organ. The embolus is the semen-injection portion of his palp.

entomologist: a person who studies insects. Occasionally, one who studies insects will also study spiders, and such a person could be referred to as an arthropodologist, but nobody uses that term. A person who studies arachnids is called an arachnologist. A person who studies spiders is an araneologist.

entomology: the study of insects and other arthropods. See "entomologist."

exoskeleton: the hard outer shell of spiders and arthropods. Instead of bones, spiders have a suit of armor consisting of a waxy cuticle, and their muscles are attached on the inside.

flagelliform: the silk gland that produces silk for the capture-spiral portion of the web.

frame: the outer border of orb-weaving spider webs made from non-sticky silk.

head: the front portion of a spider's cephalothorax where the eyes and chelicera are located.

hub: the middle part of the orb web.

invertebrate: a general term referring to any animal that does not have a backbone. Spiders, insects, worms, crabs, and octopi are all invertebrates. Most of animal life on earth has no backbone.

mygalomorph: any spider in the scientific classification of the infraorder Mygalomorphae. Only 5 percent of North American spiders north of Mexico are mygalomorphs. Unlike their counterparts, the araneomorphs, which bite in a pinching action, mygalomorphs have parallel fangs that close straight down.

pedipalp (palp): the leglike appendages extending from either side of a spider's mouthparts. Like insect antennae, pedipalps are used by spiders to smell and taste. Male spiders also use their palps to transfer sperm to female spiders.

pheromone: any one of many chemical secretions used to communicate within species. Some spiders use pheromones to communicate such messages as mate recognition.

piriform: a silk gland that produces the silk that bonds web attachment points.

spiderling: an immature ("baby") spider.

spinneret: the spider's silk-spinning organs, usually on the underside of its abdomen, near or at the very back.

spiracle: the holes, usually on the underside of a spider's body, that open to its respiratory system. Basically, it's how the spider breathes, like our mouths or noses.

spiral: the sticky, circular portion of an orb web that extends beyond the hub.

subadult: an older, immature spider nearing adulthood.

thorax: the region of the cephalothorax behind the head. The thorax has legs extending from it.

tubuliform: the silk gland that produces silk used for egg sacs.

viscid silk: the sticky, wet silk produced by aggregate silk glands.